ABOLISH SILICON VALLEY

ABOLISH SILICON VALLEY:

HOW TO LIBERATE TECHNOLOGY FROM CAPITALISM

Wendy Liu

Published by Repeater Books

An imprint of Watkins Media Ltd

Unit 11Shepperton House

89-93 Shepperton Road

London

N1 3DF

United Kingdom

www.repeaterbooks.com

A Repeater Books paperback original 2020

2

Distributed in the United States by Random House, Inc., New York.

ISBN: 9781912248704

Ebook ISBN: 9781912248711

Printed and bound in the United Kingdom by TJ International Ltd

CONTENTS

This book is memoir. It reflects the author's present recollections of experiences over time. Some names and characteristics have been changed, some events have been compressed, and some dialogue has been recreated.

ZERO: PROLOGUE

Opposing tech innovation is punishing the poor by slowing the process by which they get things previously only affordable to the rich.
— Marc Andreessen, co-founder of Silicon Valley venture capital firm Andreessen Horowitz, in a tweet posted on June 4, 2014 (since deleted)

Silicon Valley is more than a region in northern California that has become synonymous with the high-tech industry. It is a dream.

It is the dream of a world with new rules and new rulers, based on the principles of meritocracy and efficiency and hacking your way to the top. It is the dream of the win-win: innovation that generates profit through frictionless experiences and synergistic efficiencies, not exploitation. It is the dream that a hacker playground with unimaginable wealth and minimal outside supervision is indisputably making the world a better place.

But what may be a dream for a select few is steadily becoming a nightmare for everyone else.

There was a time when I believed in Silicon Valley unquestionably. As a teenager, I poured my free time into honing my programming skills and soaking up hacker culture; my idols were startup founders and open source programmers. Even as I slowly came to terms with the industry's numerous problems — how toxic it could be for women and minorities, the absurd levels of wealth directed at spurious causes — I still saw it as deserving of its vaunted role. I genuinely believed that the industry was doing good in the world.

Soon, my belief in it had become part of my identity. After university, I plunged myself into a startup in the hope of attaining some nebulous idea of Silicon Valley success. The use cases of our product bored me, but the technical problem was captivating, and I directed all my energy toward dealing with the numerous technical fires. I was confident that we would eventually succeed and then everything would be worth it.

And then one day I was no longer sure. My certainty in my personal trajectory faded, first slowly and then all at once. The fires still raged, but putting them out no longer felt pressing. The startup I had devoted my waking hours to collapsed in a blaze of mediocrity. I watched it burn, smoulder. The remnants of the misguided dream I'd been chasing crumbled into ashes.

In the meantime, the rest of the world was crumbling, too. Donald Trump had just been elected President of the United States, and in my shocked dismay I made an effort to catch up on everything I hadn't noticed while submerged in my failing startup. Everywhere I looked, things seemed dark. One ecological catastrophe after another. A housing crisis in nearly every major city. People in the richest country in the world dying because they couldn't afford insulin. All of which seemed like the morbid symptoms of a decaying socioeconomic order.

Meanwhile, the tech industry was in the spotlight for all the wrong reasons: a cascade of sexual harassment lawsuits; billion-dollar companies revealed to be built on fraud; founders climbing up the billionaire charts while their precarious workforce sank deeper into poverty. The industry I'd assumed was the solution was clearly part of the problem, and I started to see my past choices in a new light. My startup looked less like an audacious attempt to build something useful and more like an act of myopic self-aggrandisement, one that I found hard to justify once I began paying attention to the rest of the world. Creating derivative software to help brands better understand their customers seemed like a poor use of my limited time on this burning planet.

In my disillusionment, I sought out answers. I wanted to understand why everything felt like it was falling apart, and why I was only just starting to see it. At first, I didn't really know where to look — all I had were vague doubts about my worldview — but eventually, I stumbled upon an analytical framework that made sense. There were others thinking about the same questions, and their answers helped me connect the dots in a way I had never before considered. Finally I had a coherent analysis which allowed me to situate my burgeoning scepticism of the tech industry within a larger critique of capitalism.

Why had I been such a fervent believer in Silicon Valley in the first place? I could come up with all sorts of justifications, but ultimately it was because I knew Silicon Valley had a comfortable spot in the economic hierarchy. And from this soil of economic reality sprouted an entire tree of self-serving rationalisations about who was deserving and who was not, which I am only now beginning to cut down.

This is the book I wish I'd had when I started to become disillusioned with my startup, and soon, with startups, period. This is the book I wish I'd had when I began to question whether the tech industry's massive riches were deserved, and whether it was moral for me to complacently take a spot within it. This is the book I wish I'd had when I stopped believing Silicon Valley's legitimating myths — the revenge-of-the-nerds fairytales, the straight-faced claims of meritocracy, the propaganda about mission-driven companies solving the world's problems — and sought a new narrative.

This is a book about technology, but it is also a book about capitalism — that is, the downside of capitalism. Rather than treating capitalism as a wondrous force for unleashing innovation, this book focuses on the negative consequences of technological development under capitalism: the harmful tendencies that are magnified, and the alternatives that are lost.

I realise this will be a controversial perspective. The dominant narrative within Silicon Valley is that technology is inseparable from capitalism, and so innovation requires letting the free market run roughshod over every aspect of our lives. Anyone who suggests otherwise comes off as a Luddite, opposed to much-needed progress due to malice or ignorance.[1] You're either in favour of the tech industry bulldozing over anyone who stands in its way or you're against innovation entirely.

This is a convenient narrative for Silicon Valley, which then gets to frame itself as synonymous with progress: Silicon Valley is the harbinger of the future, and anyone who opposes it is stuck in the past. But the choice of being pro-tech industry or anti-tech is a false dilemma. The tech industry in its current form — with billion-dollar corporations, venture capitalists, and a few boy geniuses running the show — is not the only way of developing technology. In fact, the present industrial model is a betrayal of the liberating possibilities of technology, as technology that should serve the public good is instead locked up within corporations for private gain. Whatever progress is represented by Silicon Valley, it's certainly not the sort we should all be rooting for.

I'm not anti-tech; my critique of the tech industry stems from a place of love. I respect the craft that goes into building the industry's products, and I'm grateful for what the industry has given me. But I also believe that technology has greater potential than the mundane, profit-seeking ventures to which it has been relegated under the current system. The problem is not technology, per se: the problem is capitalism. Lurking beneath the stories about unethical products, inflated valuations, and toxic work environments are deeper structural forces that constrain the industry's possibilities, so that even the most well-meaning executives end up making decisions that harm the people they claim to be serving. Within this structure, technology that could have genuinely helped people is instead an accelerant for capitalism's most destructive tendencies.

Abolishing Silicon Valley doesn't mean *halting* the development of technology. It means devising a new way to develop technology which fulfils technology's transformative potential. It means liberating technology from the clutches of a mindless system whose primary aim is profit. It means creating a world where technology is developed according to different values, for different goals. In short, it means developing technology outside the logic of capital.

When we think of the tech industry, it's easy to get caught up in the quirks of the individuals involved. The mind conjures images of Facebook CEO Mark Zuckerberg robotically testifying in front of Congress,[2] or Uber co-founder Travis Kalanick lecturing an Uber driver about personal responsibility.[3] We associate Google with its founders Larry Page and Sergey Brin; Twitter with CEO Jack Dorsey; Apple with founder Steve Jobs and current CEO

Tim Cook; Microsoft with founder Bill Gates. As a result, we're tempted to believe that the problems of the industry originate with the specific individuals at the helm, who are somehow uniquely bad, and that if we simply swapped in some people with better ethical codes, everything would be fixed.

Of course, this instinct isn't entirely misguided; more ethical founders would certainly be better than the status quo. But the problems go deeper than the idiosyncrasies of individuals. There's a cold underlying rationality which corrupts the motives of even the most well-meaning, and in the absence of strong accountability mechanisms, individual ethics can only go so far. Ultimately the problem is structural, and the solutions will need to be structural, too.

To understand why Silicon Valley has become the way it is, we need to investigate the structural factors that explain its inner workings. We need to see Silicon Valley not as a unique aberration, but instead as the logical expression of technological development under capitalism: technology as a means to further capitalist ends. We need to consider the culture, the incentives, and the forces that explain why people behave the way they do, and why certain people are elevated over others. That means understanding the historical factors that laid the groundwork for Silicon Valley in the first place. Ultimately, the story of Silicon Valley is inherently entwined with the mechanics of contemporary capitalism.

Capitalism is one of those words with a highly contested definition: how you define it is a function of your stance toward it. So even venturing a definition requires navigating tricky political terrain. For my analysis, I'll take as a starting point a straightforward definition: a mode of production in which actors are driven by the accumulation of capital, which is made possible through private ownership of the means of production.

Of course, that definition is very incomplete; capitalism wasn't formed in a vacuum. The system we have now has developed over the course of history, sprouting over a field already littered with other systems of governance. The inception of capitalism is interlaced with various historical systems of social control, and so it's useful to understand capitalism as more than simply an *economic* system. My personal favourite definition of capitalism comes from critical theorist Nancy Fraser, who describes it as an "institutionalised social order"[4] that governs not only the accumulation of capital itself, but also the noneconomic background conditions which make capital accumulation possible. Capitalism requires a particular social arrangement which takes advantage of divisions that predate capitalism, including those based on race, gender, geography, and other accidents of birth.

On top of these social divisions, capitalism's very axioms lead to an overarching economic division. Capitalism is fundamentally a system of distinct classes[5]: those who own capital, and those who own nothing except their ability to work; those with power, and those who live under the power of others. After all, owning the means of production doesn't mean much unless there are people willing to work for those owners — even if the profit they produce accrues to the owners, not themselves. In practice this divide is usually fuzzy, but as the gulf of wealth inequality widens, the split between owners and workers sharpens.

This fissure sheds light on how technology is used in a capitalist system. In the production process, technology is a means to improve efficiency by increasing the output per worker. As long as capital has the upper hand, that means technology will predominantly be used to increase profits while minimising labour costs. Whether it's used in the process of creating other products or a product in its own right, the technology still belongs to *capital*, and it will be designed and deployed according to capitalist aims.

The more advanced the technology, and the more money behind it, the more it begins to seep into everyday life. Soon every single interaction is at risk of being controlled by whichever tech company happens to own the relevant technology. The result is a societal fissure. On one side are the people benefiting from technology's creeping advancement — the ones developing it, managing it, investing in it — and on the other side are those whose lives are increasingly dictated by privately-owned algorithms. And the more technological development is entwined with capital, the deeper the divide between the ruling class and those who serve them, and the more threadbare the avenues of upward mobility — all because that benefits the ruling class.

What this means is that there is no universal subject when it comes to understanding Silicon Valley. Silicon Valley is often referred to as a dystopia, but dystopia is a *subjective* descriptor; what is dystopia for some may well be another's utopia. A society where most people are subject to the whims of unimaginably powerful technology corporations isn't so bad if you're an executive at one of those corporations. No diagnosis of Silicon Valley can be written without this fundamental divide in mind.

This book is an attempt to unpack the problems of the tech industry, as told through my personal story of disillusionment. It follows the arc of how I got here: from unwittingly inhaling Silicon Valley ideology as a teenager, to working in the industry, to launching my own startup. And the more I learned about the industry — its internal mechanics, and its impact on the rest of the world — the more I began to doubt. In the process of searching for an explanation, I found a golden thread of connections I had never seen before, and slowly it all started to crystallise into a coherent whole. Everything clicked in a way that I could never have imagined before. I don't claim to have all the answers, but I have found a theoretical framework that helps me understand the gap between the world I live in and the world I want to live in.

The goal of this book is to piece together what I've learned. Others in the industry will have had some experiences similar to mine, and I hope they will recognise bits of themselves in a way that connects their disillusionment to something bigger. But this story is ultimately mine, and I'm not trying to speak for anyone else.

If you're looking for a Silicon Valley exposé, you may be disappointed: this isn't meant as an industry tell-all filled with famous names and salacious gossip. This is a story of failure, not success, and a very mundane sort of failure at that. But failure is not always a bad thing. The failure of my startup is why I sought out answers in the first place, and it's why I was receptive to new explanations, no matter how painful they were to hear. Failure is what allowed me to realise the folly of the amorphous goal I'd been chasing all this time.

If you love the industry and believe all the criticism to be unfounded, you won't like this book. My conclusions will probably infuriate you. But you should read it anyway, if only to understand why others might disagree with you. The golden ages of Silicon Valley are over; there's a dawning suspicion that the story Silicon Valley spins about itself is a one-sided tale of the sort told by the victors. If you've rationalised away all the bad press because of some myths you've internalised about progress or efficiency, there's only so much I can say to convince you otherwise. Ideology starts as a story but soon hardens into a shield, and only you can decide if you want to let it down.

This is not a book with easy answers; I don't have a twelve-step process for how to fix the myriad problems associated with the tech industry. The industry I want to see will require massive systemic changes, some outside the scope of what most of us can imagine today. But that's precisely the point of this book: to conjure a world that we can't even imagine right now because we're trapped in our current dead end. It's a call for a paradigm shift, in order to envision a radically better world than what we have now.

Abolition, after all, means more than mere destruction. The point of abolishing Silicon Valley is to build something better in its place.

Those who will rally in defence of the status quo will say that this is the best we can do. They'll argue that the system works fine and that major changes might actually make things worse. But that seems eerily similar to what large companies always say right before they get disrupted.

If there's one good thing about Silicon Valley as it stands today, it lies in the industry's seemingly boundless desire to question assumptions. Unfortunately, this desire is predominantly being captured by the monotonous pursuit of profit, which results in disruption *for the wrong reasons*. Maybe this desire needs to be turned in a different direction — toward the socioeconomic system that gave Silicon Valley so much power in the first place, in order to disrupt those who assumed that they would always be the ones doing the disrupting. Time to aim the cleansing fires of disruption toward Silicon Valley itself.

ONE: NO GIRLS ON THE INTERNET

We need to stop assuming that gender gaps imply sexism.
— James Damore, a software engineer who was fired from Google in 2017 after posting an internal memo criticising Google's efforts to recruit more women as "unfair and divisive"; at the time, women made up 20% of the company's technical staff.[1]

I was twelve years old the first time I had an inkling of technology's transformative potential.

One afternoon, a former classmate told me over MSN Messenger that he was building a website. It was a kind of puzzle game viewable in a web browser, modelled after a popular game called notpr0n. The premise was simple: each level was a logic puzzle, often requiring technical knowledge of how websites work, and once you figured out the answer, you would change the URL to get to the next level. It was surprisingly addictive, and when I got stuck and asked my friend for help, he told me to view the source code. I had to look up that term on Ask Jeeves, but once I figured out how to pull up the angle brackets and beautifully nested indentation, it was as if a hole in the universe had just opened up to reveal its inner mechanics. I could see the correspondence between this monospace text and the graphics that appeared in my web browser.

Before this, I hadn't realised that websites were something that *I* could make. At that point, I mostly used my computer for playing role-playing games and, occasionally, typing up the entirety of various library books so I could have electronic copies at home. But after that fateful first glimpse of a website's source code, it occurred to me that my computer could do more than *use* programs written by other people. I sped through all sixteen existing levels of the game and offered to help my friend come up with more.

By the time I turned thirteen, I had pretty much taken over the game, having added dozens of new levels, drawing on my newfound understanding of how to build websites. The scope of the game expanded, too — I started a bulletin board where players from all over the world could chat, sometimes about the latest level, sometimes just about their lives. I pored over technical documentation on how to build websites; I soaked up manuals about hacker culture; I memorised guides on good practice and forced myself to unlearn the bad habits I'd picked up. It was a slow and tortuous process, but I couldn't get enough. My after-school hours were spent immersed in text editors and FTP programs and spiral-bound notebooks where I brainstormed websites to build. I was captivated by the potential of this new medium, and it struck me that I could live my life in this virtual world, instead of the messy real world in which I increasingly felt like I did not belong.

One ambitious plan featured a discussion board for my imaginary interlocutors to discuss the state of the universe. (It was literally called "The State of the Universe"; the logo featured a blurry photo of the Earth against a backdrop of stars.) I spent months modifying the discussion board software so that it would have the features I wanted, then pre-populated the board with categories that only a severely sheltered teenager would think representative of the state of the universe. The board died before it opened — after months of extensive software changes, I realised that my ambitions exceeded my capabilities, and the board was quietly scrubbed from my Internet history.

By this point, I had a new idea: the Online Database for Media Information (ODFMI), which I conceived of as the all-purpose alternative to the Internet Movie Database (IMDb). I began the project with an almost comical ignorance of contemporary copyright law; my plan was to crowdsource an electronic copy of every book, film, or other media product ever made, seeded with the library books I'd typed up over the years. I even registered a PayPal account under that name, so confident was I in the eventual success of the project. Once I understood that my vision was impracticable (and almost certainly illegal), I settled for a database of public domain books taken from Project Gutenberg, with chapters formatted as discussion board posts. Eventually, I realised that most websites on the Internet were run by corporations, not teenagers stealing time on the family computer, and that IMDb itself had been acquired by Amazon for $55m years before. With regret, I abandoned my audacious plans to revolutionise the media industry.

Even as a pre-teen with a negligible understanding of societal gender dynamics, it didn't take me long to realise that the parts of the Internet I liked were not always welcoming to women. Beyond the cliché that there were no girls on the Internet, there was no shortage of jokes about women's place in society, or casual references to women's lack of aptitude for programming. I figured that I would simply get used to it; I derived enough joy from Internet culture to remain steadfast in my belief that I belonged anyway. Plus, the nickname I'd adopted for most online interactions was sufficiently androgynous that some people even addressed me as "Sir".

I had no guidebook on how to navigate gender politics in that thorny landscape. All I knew was that women were denigrated in this cultural space, which led me to conclude that being visibly female could be a hindrance. I wasn't sure exactly why this was the case, but my intuition told me that the poor treatment of women was warranted — women simply must have been inferior in some way.

For some reason, this conclusion didn't negatively impact my own self-esteem. Rather than seeing it as a repudiation of myself, I instead saw it as an indictment of those who shared my gender. I didn't have to be like *them*; I could transcend the physical impediment of being part of the clearly lesser half of the human population. And if I couldn't literally escape my body, then at least I could seek refuge in the mind. Through preserving my achievements on the Internet, I could attain a kind of virtual immortality denied to those grounded solely in the flesh.

Even when my ambitious website plans evaporated, my interest in the technology behind the discussion boards remained. To make the websites, I had to learn to use a piece of software called phpBB, and in the process I stumbled upon a whole new world — unlike the basic websites I had made before, these were written using a programming language called PHP, which meant they could be *dynamic*. As I pieced together how PHP worked through trial and error, it dawned on me that this discussion board software was merely one possible application of the techniques I was learning. I felt like I was nibbling at the edges of an elegant tapestry, stitched together with a coldly beautiful logic that I had never come across in my formal education. Maybe my after-school hobby could turn into something more profound.

What began as an attempt to alter phpBB to build specific websites soon morphed into an interest in the technology for its own sake. I started hanging out in the discussion boards for the phpBB project, initially asking others for advice and eventually giving out advice myself. After a few years, I became an official team member, first as a moderator and then helping to update the website. I was so excited to be part of something bigger than myself that it became my main focus, siphoning energy from school and whatever else I should have been doing at that age. Sometimes I would play truant with a dubious illness-related excuse in order to catch up on my discussion board duties, despite not being paid to do any of it.

Not being paid was part of the charm. The project was entirely volunteer-run, and its culture was planted firmly in the burgeoning open source software movement. At first, I didn't really know what that meant — I just knew that I didn't have to pay to use the software — but gradually I understood that it was more than a pricing model: it was a cultural movement. Soon enough, I fell down the open source rabbit hole, devouring whatever books and essays I could get my hands on: Richard Stallman's *Free Software, Free Society;* Neal Stephenson's *In the Beginning Was the Command Line;* Eric S. Raymond's *The Cathedral and the Bazaar;* Lawrence Lessig's *Free Culture.* I scrutinised dictionaries of jargon so I could decipher the numerous flame wars debating the finer points of GPLv2 vs GPLv3, vim vs Emacs, tabs vs spaces. And I decided to embrace open source in my personal life, starting with my operating system. Windows was insecure, and macOS was proprietary, but Linux was the primary choice of the kind of programmer I was trying to emulate.

It wasn't easy to buy a computer with Linux pre-installed, so my first challenge was to install it myself. After months of haranguing, I finally convinced my family that I needed my own computer, and my first laptop was a budget fifteen-incher picked up during a holiday sale. I loved it. I spent an entire weekend hunched on the ground with the laptop plugged into my modem, meticulously scrolling through forum posts addressing common Wi-Fi driver problems. When I emerged on Monday with a functional install of Ubuntu 7.10, I felt like I had made it through a wilderness. Once I had my own computer, there were no limits on how late I could stay up coding; the nights grew longer and then shorter, and the outside world was cold and unwelcoming, but I was always inside where it was warm, a solitary figure lit by the weary flicker of an LCD screen.

The more time I spent engrossed in this virtual culture — absorbed through IRC quotes, email archives, and discussion boards with varying levels of anonymity — the more I fell in love with it. Still, I was constantly dogged by the uneasy feeling that I was joining a space that wasn't meant for me. I could continue pretending to be male, but I didn't like deceiving the people I thought of as my community, especially since a few others on the phpBB team were openly female. And yet being openly female on the Internet seemed like a magnet for being mocked, or hit on, or both.

All my technical role models were male, and I knew of very few successful female programmers. The ones I did come across evidently made inviting targets for disgruntled critics, who would scrutinise these women's blog posts and résumé and appearance with a gravity that seemed disproportionate to me. Any women who expressed even mild criticism of the male-dominated culture were derided as Feminazis who had probably slept their way into the industry anyway. I didn't want to end up an object of ridicule whose success was deemed unearned by angry men on anonymous message boards, so I carefully studied all the apparently incorrect ways for women to behave to ensure that I would model only the correct ones. The conclusion I had internalised was that feminine traits were inferior; I coped with the ensuing cognitive dissonance by reminding myself that I didn't have to be like the others — I could be better. I knew the game was rigged, but I still thought I could win.

Yet my place in that world was clearly precarious, so I overcompensated by clinging on to what little I had. When applying to university, I picked my major with gender concerns at the top of my mind; I was swayed towards physics because it was the degree with the lowest female enrolment, which I interpreted as a sign of intellectual rigor and therefore worth. I scorned anything female-dominated, like the humanities, which I saw as a scholastic consolation prize for those not smart enough for STEM. I even applied to one school I had little desire to attend, solely because of its severely skewed gender ratio; somehow I twisted that into a sign of intellectual calibre.

Though I had applied for physics everywhere else, I ended up attending McGill University in Canada, which offered a joint degree in physics and computer science. My decision seemed validated when, upon arriving late to an orientation session, I noticed that I was the only girl; one student looked at me in feigned shock and asked, "There are girls in this program?" I treated this as a badge of honour. Clearly I was on the right path.

Despite knowing how sexist the culture was, despite observing countless instances of casual cruelty toward women — for a long time, it didn't strike me as a problem. Even as I grew out of my solitary phase and started to develop a life beyond my computer screen, I still saw Internet culture as a refuge. I didn't have the context or the vocabulary to understand why it might be problematic; I disliked those who complained about it more than I disliked those who were actively making it toxic. Whatever toxicity there might have been, I saw as an obstacle that I could overcome, and if others could not do the same, then the fault was theirs. I never questioned the legitimacy of the implicit gender hierarchy, because I assumed that I would find a way to be on top.

I always knew that there was a prize at the top of that hierarchy. Coding was clearly lucrative in a way that my other skills were not. As a teenager I would draw handmade greeting cards for my mother, who would sell them to her friends and give me a cut of the proceeds; an eight-hour session would net me approximately eight Canadian dollars and a severe aversion to the smell of markers. But even as an amateur coder prone to undercharging, I could make a hundred dollars in a day doing freelance work over the Internet. Before I was even legally allowed to work a real job, I learned that if I could make some basic changes to someone's website, then money would magically appear in my PayPal account. I never even had to talk to anyone on the phone, much less IRL.

In the early days of diving into this world, I occasionally picked up hints about the money in the broader tech industry — the land of full-time jobs and stock options. One well-known blogger wrote about sending limousines to fetch interview candidates[2]; another peppered his essays with references to the life-changing money he made when his startup got acquired. Programming was clearly more than a teenage hobby — it was a real profession, and a rewarding one at that.

At that point, money seemed to me like a fairly objective measure of merit. People who were making a lot of money were clearly being rewarded for something — because their talents were deemed valuable in some deeper sense than the financial. So I interpreted the apparently vast amounts of money floating around the tech industry as proof of my own diligence and worth. I saw it as validation for spending so much time indoors, manipulating bits on my computer, instead of doing whatever other people — less diligent, less worthy — were doing out in the sunshine. I believed that I had made the right choices, and the obverse was that everybody else had simply made the *wrong* choices, to be confirmed when they inevitably failed to secure high-paying jobs.

Programming, though, seemed like more than merely a *job*. I pictured it as a stepladder leading to a higher plane of existence, one where the nerds would get their revenge by becoming rich and powerful. The high salaries and generous perks I took at face value — as proof that I would get what I deserved. (It hadn't yet occurred to me that tech companies expected to get more out of an interview candidate than it cost to hire a limousine.)

My first semester at university was the first time I was exposed to formal education in computer science. Much of what I was learning was new to me — if not the concepts, then at least the terminology or explanations — but even then, I affected an air of yawning world-weariness. My pre-university programming experience, as minimal as it was, gave me a license to feel superior to many of my classmates, whose dearth of real-world experience I attributed to a lack of innate interest. They were merely memorising definitions to pass exams and get jobs, as opposed to the more respectable path of painstakingly deciphering unreadable code out of the sheer joy of it. They weren't *real* programmers; they didn't go through the *desert*. (I needed to believe in the inherent superiority of my journey into that world, because I didn't know how else to find self-worth.)

In my second semester, I got my first formal job as a web developer for a research lab at my university. Embarrassingly, I flubbed the interview — I had little experience with either of the languages I was supposed to be using, and the ensuing half-hour of staring blankly at a chunk of code I was supposed to fix was met with disappointment from the interviewers. They hired me anyway, presumably on the basis of my open source experience rather than demonstrated ability. I was eager to prove myself, so even outside work hours I honed my skills, cobbling knowledge together from StackOverflow, technical documentation, and, when all else failed, a strategic use of Google. It was thrilling, and I genuinely looked forward to coming in to work. I didn't even mind that the pay was barely above minimum wage — I treated the job primarily as a learning environment, where I got to develop skills I wanted to develop anyway; the money was a bonus.

The job turned out to be a godsend. My university was a large public institution whose stellar international reputation rested on its research output rather than its quality of instruction — in the computer science department, curricula were often hopelessly out of date. At work, I got my first taste of what it was like to work on an actual team building real software, and I couldn't get enough of it. It was almost astounding that this was something I could get paid to do. Even better, it turned out that our software was to be released under an open license. This was an unexpected open source refuge, made possible as part of publicly-funded research, but such refuges were rare; most programming jobs entailed working on proprietary software, or at least having their salaries indirectly funded by proprietary software. I was finding it difficult to reconcile my appreciation for the money sloshing around tech with my continued interest in the open source movement, and I was worried that there might be an irreconcilable contradiction between the two.

For the first two years of college, I continued to volunteer with the phpBB project, providing support, sharing my own code modifications, and occasionally charging money for private jobs. As with other open source projects, money was a constant topic of debate in our community. What was the right way to fund open source software when companies with proprietary technology were starting to dominate software production? How could a volunteer-run project like ours compete?

At the time, our funding model relied primarily on unobtrusive banner advertising for paid hosting. Even that was controversial; some team members thought our project should be funded solely through non-commercial donations. Impassioned arguments frequently erupted over the ethics of advertising, how people should be paid to write code released under an open license, and whether it made sense to give our code away for free when our competitors were charging. The realists wanted to explore new revenue models so we could pay developers to improve the codebase; the idealists saw making money as immoral and argued that developers should be motivated to work on the project because they believed in open source, not because they were getting paid.

When I first joined the project, the realists had seemed to me like sellouts. I had been firmly in the idealist camp: open source developers should do it out of love, not for the money. (I had been living at home, with an allowance.) But over time, I realised that my stance, however principled, didn't quite work in reality, and the contradictions confused me and left me demotivated.

For two years in a row, I travelled with other phpBB team members to Portland for a conference about open source software.[3] The first year, I had been elated to be around all these other people who loved open source, and proud to be representing a project I loved. The second year, though, there was a marked lurch towards the corporate end of the spectrum: the open source projects had been shunted to a smaller area, lunches were hosted by proprietary software companies, and Microsoft had pitched a giant tent in the middle of the exposition hall. Open source was clearly losing to corporate giants. It didn't feel good to be on the losing side, especially given that I had no idea *why* we were losing. Soon after that, I quit my volunteer role and accepted that I would probably have to work on proprietary software if I wanted to make a living in this industry.

McGill University was a hotbed for radical student activism while I was there, but it almost completely passed me by — I had little interest in politics, because I didn't see how it affected me. In 2011, I voted in my first national elections; I knew almost nothing about Canadian politics, but I'd heard that some of the New Democratic Party candidates were fellow students, so whatever, I ticked the NDP boxes in the booth. My candidates won in the apparently historic orange wave of a social democratic party forcing out the nationalist party, and I was happy that I chose correctly, but didn't really care beyond that.

The next year, my university was the site of vigorous student protests in response to a planned provincewide tuition hike, from just over $2,000 a year to almost $4,000 by 2018.[4] At the time, Quebec had the lowest tuition fees in Canada, so I found the proposed fee increase to be eminently reasonable. But an alarming proportion of the campus seemed to disagree with me: fellow students blocked classroom doors and marched angrily in the streets with red squares pinned to their shirts.

Most of my social circle was against the protests. A friend in STEM wrote a long Facebook post condemning the activists for being unreasonable, reminding them that even with the increase, fees would still be low enough that it would be possible to earn an entire year's tuition in one summer of full-time work. Another posted a photo of herself flipping off protesters as they passed underneath her balcony. I refrained from posting anything myself, but my attitude towards the protests was primarily scorn; I rather liked the idea of higher tuition fees. Not only did raising fees seem like a pragmatic policy given Quebec's fiscal situation, but it fit my worldview. I thought the move would reveal the uselessness of degrees that wouldn't lead to high-paying jobs, serving as yet another validation of my having picked a fiscally responsible major.

Not that I needed more validation. During a particularly bleak stretch of my adolescence — marked by the perilous blend of impeccable grades and lonely lunches — I had fallen in love with Ayn Rand's novel *The Fountainhead*, which I took as confirmation that I would eventually be avenged for my current lack of recognition. Around the same time, I had discovered Paul Graham, a successful programmer, startup founder and investor in Silicon Valley whose essays about nerds I read as proof that I was special and would eventually get my just deserts. And during freshman orientation, booze-soaked students in my faculty would gleefully recite chants about the inferiority of arts majors, whose career prospects were summarised as flipping burgers or pouring coffee — the ability to find a high-paying job evidently being a reflection of personal superiority. I, of course, would find a job that paid me a lot of money.

TWO: GOOGLEYNESS

We have a strong commitment to our users worldwide, their communities, the web sites in our network, our advertisers, our investors, and of course our employees. Sergey and I, and the team will do our best to make Google a long-term success and the world a better place.
— Larry Page, co-founder of Google, in Google's 2004 IPO letter[1]

By the start of my third year of university, I was doing part-time software development in two research labs while maintaining several personal projects on the side. This was on top of a full-time course load, and so I occasionally ended up pulling surreptitious all-nighters on campus to catch up on homework. A colleague who had caught me napping in our lab told me that I was working way too hard for an undergrad, and asked if I had ever considered applying for an internship at Google.

Most of what I knew about Google came from *Business Insider* listicles praising its workplace culture; my impression of Google was that it was where smart people could get paid tons of money to make the world a better place. I was hesitant at first, unsure whether I was good enough yet, but I sent over my resume anyway, and my colleague forwarded it to a friend of his who was willing to submit an internal referral for me. A short while later, a recruiter reached out to schedule a coding interview over the phone.

I read everything I could about the Google interview process, brushing up on notes from my algorithms classes and scrutinising explanations of what it meant to be "Googley". As a result, the first phone interview was a breeze: forty-five minutes of writing code in a Google Doc while strategically narrating my thought process out loud, like all the blog posts suggested. Not long after, my recruiter congratulated me on passing the phone screen; I would henceforth be in a "pool" with the other interns, waiting for a full-time employee to pick me for an interview with their team.

In the meantime, I diversified. A Microsoft recruiter came to my campus careers centre for whiteboard interviews, and a simple question about binary trees eluded me in the moment, but I improvised well enough to merit an all-expenses-paid trip to Mountain View for another interview. I had zero interest in working for Microsoft, which I saw as the villain in the free software wars, but a free trip to California was hard to refuse in the depths of a Montreal winter. At Microsoft's Mountain View office, after a series of further whiteboard questions, a disinterested employee asked me why I wanted to work on Outlook.com, and I stammered something unconvincing about a fun learning experience before trailing off into silence. I did not get an offer from Microsoft.

Facebook was another option. I was never especially keen on their product, but they seemed to be in the same engineering calibre as Google, and that was all that really mattered. A friend who had already interviewed with them connected me with a recruiter, and I conducted another forty-five-minute phone interview, this time not in Google Docs. My first question was nearly identical to the one I flubbed for Microsoft, and somehow, I nearly flubbed it again — I could not believe that I was simultaneously so lucky and also so stupid — but because the interview was over the phone, I had time to stealthily look up the answer in my notes and pass that off as something I had just remembered, and the interviewer was happy.

In the meantime, after several anxious weeks of not hearing anything from Google, I got matched with three teams at Google in one week. One team worked on infrastructure management, which sounded intriguing, and they were based in San Francisco rather than the Mountain View headquarters. My imagination filled with visions of charming hills and cable cars and biking across a red bridge in the low sun. Before I heard back from Facebook, and before I even talked to the team, I was sold on the basis of the location alone. Facebook's offer turned out to be similar, but Facebook would have me work in their Menlo Park office, about an hour's drive south of San Francisco, and that seemed much less glamorous. Anyway, I had already signed my Google offer.

Both companies offered shockingly good pay, especially compared to my part-time wages of $15 CAD an hour. After running some basic calculations, I concluded that I would make about $20,000 USD that summer after taxes, with transportation and housing all paid for by my employer. This was more than my mother — who had a master's degree — earned in a year. As a result, my initial reaction was to feel horrifically grateful, because I didn't know why I deserved this. But in my head, I was already crafting a narrative to fit the facts: I did deserve it. I had worked hard, and now I was reaping the rewards.

Google's internal PR efforts did nothing to dissuade me of my burgeoning conviction that I was unusually deserving. In May of 2013, they put me on a nearly empty direct flight to San Francisco where the person sitting nearest to me also happened to be a Google intern, and my lodging for the next few months was a two-bedroom luxury apartment on Beale Street, a ten-minute walk from the Spear Street office. I had to share a room with another intern, but the apartment was nicer than any place I had ever lived, and I was scared to look up how much it cost.

The first week was intern orientation. That happened down in Mountain View, so in the mornings I boarded a private shuttle that took me down the 101 for the hour-long commute. I was mildly carsick, but as the bus followed the gently curving highway out of the city, the fog began to lift, and for a moment I was lost in the unimaginable beauty of this place, sunlight glinting off the limpid blue surface of the bay to my left, the soft tendrils of fog that circled the hills to my right dissipating in the morning light. I didn't ever want to leave.

The Mountain View office felt more like a university than a corporate headquarters, so it wasn't surprising that everyone called it a campus. The grounds were filled with assorted colourful amenities: human-sized robot mascots, branded bikes lying unlocked on the grass, cafeteria tables festooned with sun umbrellas. In a bright room crammed with fellow eager interns, in between cheery paeans to how wonderful Google was, I signed a long-winded employment contract I assumed they didn't actually expect us to read. Recruiters dazzled us with statistics about the amount of data Google used and showed us sentimental marketing videos about the amazing things Google was doing for the world. Most importantly, we were told that it was a privilege for us to be here, inside this wonderful and fast-growing corporation, and so we should treat it like a family. We would have access to lots of secret data, but the flip side was that it had to stay secret; if we shared anything unauthorised, Google would know. After all, Google *did* own Gmail.

We were taught the corporate motto: organising the world's information and making it universally accessible and useful, which I liked because it reminded me of my ill-fated attempt at creating a database for media information. But the sensibleness of this motto made the cloak-and-dagger stuff seem a little absurd. What's more, it wasn't clear to me that organising the world's information should be the provenance of a for-profit company, rather than a democratic effort outside the purview of the profit motive. Still, it was nice to get paid so well, so I couldn't really complain.

The week capped off with an internal town hall-style event called TGIF[2]. Every Thursday at 4:30pm, founders Larry Page and Sergey Brin would get up on stage in an auditorium in Mountain View to share company news, answer employee questions, and welcome new hires. This week was the beginning of intern season, and a few dozen of us waited quietly to the side; we'd each been given a grab-bag of "Noogler" — New Googler — swag, including a t-shirt with our name and role embroidered on the sleeve. I eyed a propeller cap decked out in the Google colours with distaste — we were supposed to wear this? We were interns, but we weren't children. I put it on anyway and begrudgingly stood up with the other Nooglers as the audience dutifully clapped.

It was a little weird, I had to admit, a little cult-like. But the consensus on Google was that it was a great company to work for, the best of all corporate jobs. I should shut up and enjoy the summer.

Which wasn't hard, as it turned out: the company had clearly put effort into making their internships as enjoyable as possible. There were after-work intern activities all the time: a margarita-making class for the over-21s, a kayaking trip around the bay, a party aboard a boat. And we got to attend the weekly TGIF sessions, which were a delight even beyond the free alcohol and snacks served. One Thursday, during an announcement for some cooking-related product, the stage was decorated with kitchen counters covered in assorted produce and appliances; Sergey periodically got up for some improviso puttering around the kitchen, and Larry affected a comical annoyance every time the blender whirred while he was trying to talk. A fellow intern and I looked at each other, like: seriously? We joked that Sergey should stop drinking at work. It was a little bizarre, but it was fun. You couldn't ask for more approachable founders.

For the San Francisco Pride Parade, Google went all out — the company had a float, and employees got T-shirts with the corporate logo decked out in rainbow colours. My apartment building housed about a dozen Google interns, and we organised Pride-adjacent day drinking from 10am onward; by the time we got to the float I was completely alcohol-dissociated, my memory wiped. Later, forgotten photos of me racked up likes on Facebook, and I changed my profile picture to one of me beaming in a Google Pride T-shirt, holding a bright red Google-branded balloon. Google's progressive appearance was another reason to be proud to work there. So convenient that they did politics on my behalf — I could feel virtuous solely through my association with my employer. All I had to do was show up.

Then there were the office facilities. Adorning the entrance to the office was a multi-floor tube slide, rarely used except by the occasional delighted visitor. Near my desk was a small gym, and sometimes I would sneak off to do pull-ups while waiting for my code to compile. One evening, exploring the office after hours with some other interns, we were delighted to discover a music room: drums, guitars, microphones, video chat connected to music rooms in other offices.

For a few weeks, it was dazzling. I was touched by the thoughtfulness of this company that clearly cared about creating a comfortable environment for employees. The cafeteria on the top floor of my office opened to a terrace facing the bay, right under the Bay Bridge, and most days I woke up just in time to catch the end of breakfast, piling my plate with free-range egg whites and organic steamed kale and maple-glazed bacon. My first half hour at work was typically spent soaking up sunbeams as the morning rays carved out wisps from the fog curling underneath the bridge, feeling astonished at just how unbelievably lucky I was to be here.

Other companies in the area were equally dedicated to luring interns with free food, drinks, and expensively-arranged fun. My Facebook recruiter invited me to an intern carnival at their Menlo Park campus, and I carpooled over with other Google interns to find an entire ersatz amusement park on their manicured lawns. And payments startup Square, founded by Twitter co-founder Jack Dorsey, hosted a party for Bay Area interns featuring a full sit-down dinner that ended with cupcakes topped by an edible Square logo. Square's main product was a credit card reader that plugged into a smartphone, which I didn't find particularly exciting, but they had competitive salaries and a nice office and that was good enough for me. I snuck some photos of Dorsey and took as a souvenir some of the Square readers which littered the office as paperweights. That their office used to be the *San Francisco Chronicle* building seemed symbolic: a new industry subsuming the old. How fortuitous for me that I was part of the new.

In June, Google interns were invited to the premiere of a new film called *The Internship*. The movie, starring comedy duo Vince Vaughn and Owen Wilson, featured a fictionalised portrayal of working at Google, and it was partially filmed on the Mountain View campus with permission from Larry and Sergey. As a depiction of the Google internship experience, it was absurdly unrealistic: the Google internship was portrayed as a *Hunger Games*-esque competition, complete with Quidditch tournaments and hackathon-style all-nighters. Afterward, we all made fun of the movie for the contrived nature of the internship challenges. Of course Google engineering interns would never be asked to sign up local businesses for Google AdWords — our time was much more valuable than *that*. Still, the movie did acknowledge that Google interns were smart, hardworking, and deserving, which I was gratified to see reflected in popular culture.

The work itself turned out to be the most mundane part of the summer. My project, an internal visualisation tool, was at best a "nice to have" for a few dozen people inside the company, and as far as technical challenges went, it was a little boring. The first few weeks, I was ready to work late hours to prove my dedication, and my manager showed me an intranet webpage to turn the lights back on when they automatically shut in the evenings. But there was rarely anyone around to witness my hard work, and soon my routine became more and more leisurely: an hour every morning catching up on emails, social media, and Memegen, Google's internal meme generator which closely tracked the pulse of the company. After a few unhurried hours of coding and code review, I would usually leave before 6pm, since dinner wasn't served in the San Francisco office.

Midway through the internship, the novelty began to wear off, and what settled in its place was an inexplicable, grinding dissatisfaction. I started to resent the so-called fun amenities, the saccharine statues littered around main campus, the hordes of excited tourists taking photos as if they were at Disney World. And the commute which was so breathtaking the first week soon lost its lustre — my desk was in the San Francisco office, but half of my team was on main campus, and I had to take the bus down to Mountain View roughly once a week. The drive was too bumpy to get work done, so I mostly stared out the window at the admittedly beautiful landscape and felt kind of numb.

I was about to enter my last year of university, which meant I needed to decide what I wanted to do after graduating. The most obvious option was to work full-time at Google, and there was an interview process specifically for third-year interns who wanted to leverage their internship into a job after graduation. But after a few weeks, the thrill of the perks had worn off, and the monotony of the work had crushed my initial enthusiasm for converting to full-time. I couldn't explain why, but merely the idea of going back there felt soul-crushing. Neither was I inspired by the idea of doing a master's degree, which I saw as my next-best alternative. Of course, this raised the question of what I would not consider to be soul-crushing, but thinking about *that* felt like looking down a steep cliff at an unending fall.

The main problem was that I no longer knew what I was supposed to be optimising for. I'd spent most of my life trying to succeed at whatever academic challenges were placed in front of me under the unquestioned assumption that it would all eventually pay off. And yet, the life that I saw stretching before me through the windshield of the Google bus did not look especially enticing. I felt like I had done everything right, yet the reward seemed hollow. There had to be more than this.

Meanwhile, Google's veneer of being a dream workplace started to crack. One TGIF, Larry and Sergey publicly announced an employee's firing. We were told little except the employee's name, and that he was fired after an internal investigation concluded he was the source of a press leak. I'd read the news article that triggered his termination; it was essentially a puff piece about a new product in the works, and it didn't paint Google in a bad light. But betraying the family was betraying the family, and the punishment for that was to be terminated.

"Terminated" was the word Sergey used, anyway; he dropped it onstage multiple times, Larry quickly correcting him to "fired" each time, to audience laughter, and Memegen buzzed with jokes about the incident for days. Was this supposed to be funny? An employee got fired for a minor and relatively harmless infraction, and other employees took the side of their employer rather than their fellow worker? Soon, Memegen had moved on, the leaker promptly forgotten from company consciousness, but I was still uneasy — about the extent to which Google surveilled its own employees, about billionaire founders publicly firing a low-level employee who didn't seem to have done much wrong.

The accepted stance on leaking was that it betrayed trust, and therefore needed to be punished. After all, it was for our own good. As we were frequently reminded, Google was a haven of access and trust; not only did we get great pay and benefits, we were also the beneficiaries of a spectacularly transparent internal culture, and if employees couldn't keep the company's secrets, our benevolent founders wouldn't tell us anything at all.

The transparency thing was at least partly true: I spent untold hours at work trawling the company intranet, awestruck that as a lowly intern I had access to non-public records of Google's attempts to acquire various domain names for staggering sums of money. But there was a lot of data I didn't have access to, including the data I needed to do my job — capacity projections for hardware for which I was supposed to be writing a display interface. My manager, who had to generate fake data for me to work with, suggested that we wouldn't want vendors to know our projected server needs, as they might collude to jack up prices. That sounded a little paranoid to me, but I wasn't exactly in a position to object.

The whole secrecy act was getting to me. Most Google

buildings were fairly open to any full-time employee, but in my building, there was an area that was off-limits unless you worked on the Chrome security team. On main campus, the building with my favourite restaurant was only meant to be open to employees working on Android, though other employees would occasionally sneak in for sushi. And Google X, the "moonshot" lab, kept its plans strictly under wraps.

I understood why, of course: the point of secrecy was to ward off leaks and mitigate security risks. The latter made sense to me, but the former had me unconvinced. Why was this company so afraid of public criticism? Developing technology in a bubble seemed like a recipe for disaster when people around the world were depending on Google's products — especially when Google's technical workforce was wildly unrepresentative of the wider population.[3] And the "Don't Be Evil" motto served primarily as a reminder that Google *did* have the power to do evil, and people on the outside — or even on the inside, without the requisite level of security clearance — had to simply trust the insiders to do the right thing. But such trust seemed a little misplaced, given that even minor leaks were treated as career-ending betrayals; I didn't even want to think about the penalties for whistleblowing or publicly challenging major decisions.

It probably goes without saying that none of the software engineers I knew were in a union. What little I knew about organised labour was that unions were for miners and factory workers in previous centuries, not white-collar tech workers in the new information age. Our work wasn't the same kind of work, and I assumed that meant we didn't have the same kind of grievances — Google employee complaints were rarely more substantial than resentment over the Meatless Mondays program or the wrong brand of sparkling water in the microkitchens.

Occasionally, however, I noticed spates of incipient

workplace activism in the form of employee backlash over product changes. A few months before I started, Google announced that it would be shutting down Google Reader in the face of declining usage, and the ensuing internal uproar even had some employees volunteering to maintain Reader themselves.[4] And internal criticism of Google Plus' proposed Real Names policy — which would require users to publicly display their legal names across Google platforms — was fierce enough that the company would later backtrack.[5]

Other than that, though, the work environment seemed fairly idyllic. I rarely got the sense of any separation between the company and its employees: I never heard of disputes over pay or benefits, and the swelling tide of sexual harassment allegations didn't reach me until after I was gone. The pervasive sense of tranquility made my disillusionment harder to justify — everyone else seemed happy, so why wasn't I? The work was undemanding, the pay was great, the employee badge granted me access to free food and Wi-Fi anywhere Google had an office. My inchoate workplace disillusionment felt petty and illegitimate compared to the generosity of the perks.

But not everyone working at Google had the right kind of badge. Every office had a large contingent of red-badged contractors: hospitality workers serving food and cleaning the facilities; security staff guarding the cafeteria doors to prevent unbadged outsiders from sneaking in; bus drivers steering the commute to and from all the local offices. As ignorant as I was, with my default assumption of corporate good intent, I wasn't naive enough to assume that Google's famed transparency and perks extended to these workers, too.

Even as an intern, I was allowed to treat two guests a month to the free food and facilities. No one really cared if you went over; it was an honour system in which everyone was expected to be appropriately Googley. But these red-badged workers — did they get to bring guests? Did they get to fill take-out containers to bring home? I never saw them on the buses, and I doubted any of them lived close enough to commute by foot. In the vicinity of the Google San Francisco office there was little but startups, tourist traps, and luxury residential buildings with one-bedroom apartments renting for $4,000 a month.

At the time, the Bay Area was buzzing with debate over the potentially negative consequences of the growth of the tech sector: gentrification, rising costs of living, long-term residents getting priced out of the city. While the details eluded me, I had a feeling that Google's service workers were not reaping much benefit from the tech boom, even if they did work for what was supposed to be the best employer in the country.[6] My own experience at work might be a bit dull, but at least I was learning new skills while getting paid a lot; if I came back as a full-time employee, I would expect stock grants, performance reviews, a yearly compensation approaching half a million after a few years. Service workers, on the other hand, seemed to have a much shorter path for advancement, and a much lower starting point. Was this fair?

Part of me wanted to believe that I deserved what I got because I had worked hard, while others simply hadn't. But looking at the mostly black or Hispanic red-badged employees, I couldn't escape the possibility that there might be structural reasons for this divide, independent of individual work ethic; not everyone had the same opportunity to have their hard work fall neatly along a path of upward social mobility. These structural inequalities might not be Google's fault, but Google clearly wasn't reluctant to make use of them for the purpose of lowering labour costs.

And yet this was a multinational company with billions

of dollars in market capitalisation, frequent accolades for its mission-driven culture, and socially progressive founders. Surely it was too big to fail any crucial moral tests. I came up with rationalisations: if this were a problem, someone would have said something by now. But once I went down the rabbit hole of seeing the company as an agent with its own drives, it was hard to go back. I considered the benefits of the internship program from the company's perspective: why did Google put so much money toward keeping their summer interns happy? After all, it was a for-profit company, not merely a charity to massage the egos of those who considered themselves intelligent.

There was an easy answer, of course: the internship program was a glorified ramp for the recruiting pipeline. Google needed a steady stream of tech talent in order to maintain its dominance, and recruiting costs in the industry were not cheap. Even paying interns to play Quidditch all day would be worthwhile if some of them converted to full-time or referred their friends.

But that just pushed the question down the line. If Google did make enough from each hire to offset the high costs, where did the surplus money go? Who got to decide that allocation? And why was this system fair?

My employment contract offered little in the way of explanation — it was clearly crafted to give away the least amount of information in the most legally binding amount of text. Still, I picked out intriguing phrases like "arbitration" and "exempt". The former had something to do with resolving workplace disputes through a neutral third-party, rather than through the court system; the latter meant I wouldn't get paid overtime if I ever chose to work more than forty hours a week. I didn't really understand the implications of either; all this labour law stuff was a black box to me.

I wanted to believe that all this legalese was merely

box-ticking. This new world of technology, with its quirky founders and work that wasn't really work, transcended the archaic legal framework mandated by the government. It wasn't like I planned to sue my employer; surely that was the province of the weak, those inclined to play the victim because they knew they couldn't win otherwise. I, on the other hand, would always be loyal, and I would always be valuable.

At least, that's how I had felt at the beginning of my internship; as I got closer to the end, my inclination toward unquestioned loyalty was starting to wane.

It was August 2013, and everybody wanted to talk about Marissa Mayer's recent appearance in *Vogue*.[7] Mayer was one of a small number of high-profile women in the tech industry, going from Stanford to Google to CEO of Yahoo; *Vogue* estimated her net worth at $300m. I felt like I was supposed to find her arc inspiring, and yet whatever women-in-tech solidarity I might have had was overshadowed by our $300m difference in net worth.

That seemed like an unfathomable amount of money to me, regardless of how hard she might have worked, and it also didn't feel right from the perspective of overall economic efficiency. Someone who happened to be employee number twenty at a now multi-billion-dollar company made the leap to become CEO of a rival company; now she had *Vogue* spreads, a penthouse at the Four Seasons, cash to fling at an $8m art installation on the Bay Bridge. Meanwhile, the city's homeless population continued to grow, a problem no doubt exacerbated by rising rents from the influx of tech wealth.

Of course, this wasn't the *fault* of anyone working in tech, no matter how rich; the problems predated the industry's recent boom, and they were larger than any individual. It would be unfair to ask tech millionaires to take responsibility for a problem they didn't cause purely because they lived in the same city. But I couldn't help thinking that they could do *something*. If those with more money than they could spend in a lifetime chose not to use that money for social good, then where would the funds come from? What was the point of allowing individuals to amass such wealth if they didn't use it to help others?

The first time I walked to work, I had tried taking a shorter path than the one recommended by Google Maps. There was a flight of stairs across the street from my building, which I had thought would get me to the overpass faster. But that day, the stairs had been covered with a mysterious pink goo that smelled faintly of bubblegum; I had interpreted it as a sign that I was going the wrong way, and took the recommended route instead. By the end of the week, I had solved the mystery: it was soap. The stairs were reasonably secluded, and in a city with a large homeless population coupled with limited access to public sanitation, that meant they required frequent cleaning.

I wondered what life was like for those who never made the shortlist for the next CEO of Yahoo — the renters being priced out because tech money was flooding into their neighbourhood and their landlords had no compunction about raising rents in response, the homeless residents who didn't even have convenient access to public toilets. An art installation over the bay might be nice and all, but it felt like an odd outlet for philanthropy in a city that was clearly suffering.

The trend lines seemed like they were going the wrong way, and I didn't feel good about the prospect of returning to this painfully divided world. But I didn't know what else to do — Google was supposed to be the pinnacle of what I'd worked for. Maybe I wanted a different team, a different campus, a master's degree. I told my manager I wasn't sure

I wanted to come back; he told me to interview anyway, because it would be faster to convert via an internship than to apply from scratch.

On main campus, I breezed through some whiteboard interviews — by this point, the questions were all starting to look familiar. One interviewer tipped his fedora at me as he exited the conference room. That evening, sitting on the evening shuttle to San Francisco, I ran through my options and decided they weren't all that bad. If I regretted coming back to Google, I could always switch jobs; there were plenty of options in this city. Loose feathers of cloud hung low on the fading blue horizon outside.

THREE: MORE THAN THE MONEY

If you suppress variations in income, whether by stealing private fortunes, as feudal rulers used to do, or by taxing them away, as some modern governments have done, the result always seems to be the same. Society as a whole ends up poorer.
— Paul Graham, in his essay "Mind the Gap" from May 2004, which proposed that increasing income inequality in a modern society was a "sign of health"[1]

Going back to school in Montreal after a summer in the Bay Area felt like a let-down. It got worse when Google extended me a return offer before the semester even started: I merely had to graduate and I would have a six-figure job waiting for me in the Bay Area, not contingent on grades. Once the offer arrived in my inbox, my dedication to academia plunged, and soon afterwards I left my part-time research assistant gigs and forgot about the prospect of a master's degree. There was no need to work hard anymore; I'd already been assessed, and I passed.

The offer letter was of the exploding variety: if I didn't sign it within a short timeframe, it would expire. And typical of Google, I needed to install a special Chrome app merely in order to view it. It was the standard new graduate offer, with a signing bonus that exceeded the entirety of my tuition fees, restricted stock units, lawyers to deal with immigration paperwork. In short, it was the best offer I could have hoped for. I asked for a month-long extension of the deadline.

Why was I so indecisive? Wasn't this exactly what I had

worked for? Whenever I had to explain my hesitation about the offer, I would mumble something about Google being too stodgy; that usually got some disbelief, given that Google was portrayed in the media as a mythical land of ping-pong tables and massages, but at least it was within the realm of comprehension. The deeper reasons lurking beneath the surface, I didn't know how to bring up at all: the monotony, the drudgery, the dread that if I worked a 9-5 at a big company then that work would be all that I was. The fear that wanting more would be conceited.

Even harder to explain was my ethical aversion to Google's business practices. That wasn't something I would want to talk about except with the most hardcore free software advocates, few of whom I knew personally. One thing that had bothered me during my internship was being discouraged from working on open source projects except with permission from Google's lawyers. Alarmed, I had asked my manager for advice, because I had planned to work on side projects during the summer; he had told me not to worry, and it had turned out fine, but I didn't like the principle. If I decided to go back to a full-time role, all my work would be driven by Google's needs, and would become Google's intellectual property. Even if Google did choose to release any of my work under an open license, it would be on Google's terms, not mine.

My Google offer left pending, I started interviewing with other companies, hoping to find something that felt less ethically compromised. Early September, during an hour-long lunch interview at the Montreal office of McKinsey & Company, I got a crash course in management consulting. At the end of it, I was reeling with confusing visions of Excel spreadsheets, PowerPoint presentations, and seventy-hour workweeks before going to business school and coming back as an associate, whereupon I could dump all the grunt work onto younger analysts and focus instead on racking up frequent flyer miles. That didn't sound especially fulfilling. I crossed "management consultant", ink still drying, off my list.

One interview offer I never took up. At the beginning of the semester, I had written a blog post about a mild security vulnerability that I'd discovered in a piece of software used at my university. Unexpectedly, the post had gone viral, even climbing to the top of *Hacker News*, the popular news aggregation service created by my startup hero Paul Graham. Not long after, I got an email from the company making the software, and at first I was terrified that it would be some sort of legal threat, but it turned out to be an invitation to interview for a job. No, I responded politely, but thanks. That company did not rank highly in my mental hierarchy of tech companies, and despite my lack of enthusiasm for returning to Google, I would much rather have that on my resume.

A Canadian tech company famed for its aggressive recruiting tactics came knocking. A former classmate had started working there the previous summer for what was supposed to be an internship, and he still hadn't come back to school; by the looks of how quickly the company was growing, I figured he was more likely to retire by thirty than to finish his degree. Their recruiters went all out — putting me up in a nearby hotel, plying me with drinks, and telling me stirring tales of the exciting technical problems they were solving. I didn't really want to stay in Canada, but they seemed nice. I asked about their approach to open source software, and they responded that while their core technology was obviously proprietary, they were big fans of open source software, and they wouldn't mind me working on my own projects after hours, though their lawyers would be on the alert if I ever tried to start a competitor in my spare time. That sounded more lax than Google, but it still wasn't quite what I'd hoped for.

It wasn't like I didn't understand the rationale. I had done my research on intellectual property law; I knew why Google needed ownership over employees' work even if it was done after hours, outside company premises, on personal equipment. I knew that it would be bad for corporations if their employees were able to start competitors. I just wasn't totally convinced that what was bad for corporations was necessarily bad for society. On the other hand, if the wellbeing of corporations was not synonymous with the wellbeing of society, why the hell had I put all this energy into being a good worker for some corporation?

I needed to believe that my ingrained work ethic flowed from a higher truth. I needed the money I'd been offered to mean *more than the money* — more than merely a bribe for working on something that might give me ethical discomfort. The generous offers bestowed on me by revered corporations had to have a deeper meaning than mere tokens for purchasing commodities; what I really needed from a compensation package was external validation of my past choices. As long as I could believe in the truth of meritocracy — that the best would rise to the top, and would subsequently reap the rewards — then I would have external proof of my own worth, and everything would be fine.

Yet it was hard to shake the doubts. During my summer in San Francisco, with its clear markers of poverty in a city flush with tech money, I would feel a pang of guilt every time I called a Lyft or visited a nice restaurant. I'd previously spun myself a narrative to explain my success as deserved, but I couldn't stretch it far enough to account for the realities of inequality in San Francisco. Even those who deliberately chose not to work surely didn't deserve to starve in a city teeming with wealth.

My discomfort with the tech-exacerbated inequality in

San Francisco was starting to brush up against my latent belief in the goodness of the tech industry, and I didn't know how to reconcile the two. Having started university not long after the 2008 financial crisis, my impression of Wall Street was that it was merely a clever form of grift. One of my algebra professors devoted an entire lecture to a mathematical derivation of the Gaussian copula's role in the subprime mortgage crisis, interspersed with wry commentary on Wall Street greed. I saw the vast amounts of money flowing to *that* sector as illegitimate, with the recent financial crisis serving as proof.

Tech wasn't like that, though. Sure, there was a lot of money swimming around the industry, but everything was justified because the industry was doing good in the world. Sure, startups sometimes pursued frivolous causes, and sometimes they spectacularly imploded, but the overall impact of the industry was positive — society as a whole was better off if startup founders could become billionaires. The ample value that Silicon Valley captured was still significantly less than the value created.

But maybe people on Wall Street believed the same thing about their industry.

So maybe it was easier not to think about it too much, because the more I thought about it, the closer I got to the event horizon of a lurking black hole of self-doubt.

The semester passed by in a haze. I sleepwalked through classes in numerical analysis and computational theory. As intellectually satisfying as it was to learn about lambda calculus and deterministic finite automata, none of this stuff was helpful for dealing with my current existential dilemma involving an exploding job offer and the relationship between intellectual property law and corporate power. I felt like I was waiting for the rest of my life to start, because

I couldn't resolve any of the big burning questions while I was still on campus.

In the course of my four-year computer science degree, I never had to take a single class on anything remotely resembling ethics. None of my homework assignments were meant to help me navigate the fraught gendered and racialised dynamics of applying for jobs or negotiating salaries. And none of my textbooks explained why I should feel good about making six figures writing proprietary software for a multi-billion-dollar corporation when an increasing number of my neighbours were being made homeless.

I wouldn't expect them to, of course; it felt silly to even contemplate it. That was all politics, which was irrelevant to technology. I would have to figure this out elsewhere.

My classmates were in similar boats, simultaneously lost and gratified, lacking a definitive map to make sense of it all. We gathered piecemeal, sometimes contradictory information from blog posts, StackOverflow, Quora, Facebook groups; we swapped offer letters and debated the merits of negotiating. I heard through the grapevine about someone who successfully negotiated his new grad offer to get an extra $200,000 in stock grants. I felt like I should at least try to negotiate, but I was too cowardly to really consider it, and on some level it felt like bad form to do so when they were already offering so much money.

My family couldn't help with this. My mother, an administrative assistant at my university, didn't view her career as something that defined her the way I did, and the terms of her job were collectively bargained anyway. My father, on the other hand, was the world's firmest believer in meritocracy, as he saw his educational accomplishments as having catapulted him out of rural poverty. His main advice for me was to concentrate on my academic achievements, under the assumption that everything would flow from there — in high school, coming home with a B+ on a report card had been tantamount to treason in his eyes. The most I could hope for from him was a grudging acceptance that computer science was even a real degree; everyone he saw as an intellectual equal had a PhD in a hard science. He couldn't understand why I wasn't interested in pursuing a graduate degree that would get me a prestigious job at a pharmaceutical company.

I didn't know what to do. I didn't know who to talk to. I felt conflicted about feeling conflicted when this was clearly a frivolous problem in the first place. Some of my peers in other departments didn't have any job offers, and were considering going further into student debt so they could become more employable; I should have been grateful to have anything at all. I was angry at myself for making such a big deal out of what was clearly a gift, a prize, something to be treasured. I signed the damn offer.

The malaise didn't go away once I signed. In the absence of technical achievements to look forward to, I focused on the material aspects. I procrastinated doing homework by reading articles about living in San Francisco. I downloaded floor plans for a new luxury apartment building in South Beach. I came up with a monthly budget breakdown: savings, tax, monthly expenses, amortised one-time expenses. I knew it was a distraction, but I needed the frisson of the physical things so I didn't have to dwell on my lack of enthusiasm about the work.

I found it faintly disturbing to realise that I was looking forward to the money more than I was to the actual job. What had happened to doing it for the love? But I didn't know how else to feel — despite having high earning potential, programmers were supposed to be motivated by the challenge rather than the paycheck, and that made for a confusing relationship to money. Any job applicant who confessed that they needed the job to pay their bills would get an askance look and a "no hire" recommendation. And yet, how excited could anyone really be about data visualization for capacity planning?

That semester, fall of 2013, it happened that the nascent collegiate hackathon scene was spreading like wildfire. I didn't have anything better to do, so I soon started filling my weekends with trips to hackathons at nearby US universities like MIT and Princeton. It was a gratifying escape from school, even if it was nearly impossible to explain to outsiders why I would tolerate ten-hour bus rides, sleeping on linoleum tile, and 4am pair programming sessions in exchange for Red Bull, pizza, and startup T-shirts. Some people would go to network for job opportunities, but I already had a job offer, so I was chasing something a little more intangible. There was an unexpected camaraderie to be found at these hackathons, and every time I breathed those sparks I felt alive.

In the midst of hackathon season, the hackathon club at my university hosted a visitor who was a bit of a celebrity in our admittedly insular world. Alexis Ohanian, co-founder of the successful Y Combinator-backed startup Reddit, was doing a tour to promote his new book *Without Their Permission* — as in the startup truism "ask for forgiveness, not permission". I was awestruck to meet him and immensely envious of his success. After the talk, we took him out to a bar and did a round of shots with him, the photo proudly shared on social media afterward with the hashtag #withouttheirpermission.

Our hackathon club was putting on a hackathon of our own, modelled after the ones we had attended at the more prestigious American universities. Even though this would have been our university's inaugural hackathon on the circuit, raising money was shockingly easy. I'd helped organise events for student groups before, but this was an entirely different game: free food and transportation, no cost to attendees, sponsors happy to wire us thousands of dollars without ever meeting in person. A few weeks before the event, we realised that we were several thousand dollars short, and asked one of our sponsors for more; they coughed up the money immediately, no questions asked.

I shouldn't have been surprised — hackathon sponsorship went in the recruiting budget, the events being a notorious feeder for internships and full-time jobs. This reality created a peculiar relationship between sponsors and attendees: it was a little like how a butcher might look at a calf. Except these calves were happy with the arrangement; from their perspective, it looked like a good deal.

For me, the idea of surrendering most of my waking hours to the command of an uninspiring company did not sound promising. I was hoping, beyond reason, for an alternative.

And then it came, out of the blue. I was doing homework in the computer science building one day when Nick stopped to chat. I hadn't known him for very long, but he was easy to get along with, and we had worked on a hackathon project together. What if, he proposed, we started a company together? He had recently joined the computer science lab I used to work in, and he thought the inference system I built there could be spun out into a successful startup. And as the person who had worked on it for three years, I would be the perfect CTO. He thought we might even get into Y Combinator, the accelerator founded by my startup idol Paul Graham.

I had technically already committed to working at Google, but I was intrigued nonetheless. Even though I had devoured copious blog posts about startups, and even though I read stories about startups on *Hacker News* almost daily, it hadn't occurred to me that a startup was something I could just *do*. But it sounded like a viable alternative to the dread I associated with Google, so I looked over the application for Y Combinator and tried to come up with a clever response to the question of when I last "hacked some (non-computer) system". I doubted we would get in, but it was gratifying to fill out the application nonetheless, feeling like I was part of an esteemed tradition that included the Reddit founders.

But then, mid-April, when I was half-heartedly studying for my last exam, we heard back from Y Combinator: they wanted us to fly down to California at the end of the month for a ten-minute interview. This seemed like an ungodly amount of travel for such a short interview, but they would cover some of the costs, and of course I would never pass up the chance to interview with them.

This was the moment when everything changed. With the acceptance email in my inbox, it crystallised for me that this startup could be a real alternative to the glum path I had committed to. My knowledge of startups was mostly gleaned from breathless reports in the tech press, founder blog posts, and Neal Stephenson's 1999 novel *Cryptonomicon*; in my head, it was challenging, glamorous, perhaps even dangerous. Finally, I had a chance to be the hero of my own story. In any case, if we failed, Google would always be there.

Nick convinced the professor leading the research lab to take part in this new venture, and some other students expressed interest too. My boyfriend Liam, a software engineer who had previously planned to move back to England, decided he was all in; my friend Tim, who had accepted a full-time offer for an investment bank, offered to join if we ever needed a finance person.

The day after my last exam, we flew out to California and drove to Y Combinator's office in Mountain View. The conversation in the rented van quickly quieted as we pulled up to the bright orange sign. It looked a little gaudy among the muted green and grey of its suburban surroundings, but at that point, it was the most glorious shade of orange I had ever seen.

Through a window, we saw a row of people sitting at a table. Was that Paul Graham? And next to him was Sam Altman, the newly appointed president of the accelerator! We checked in at the front desk and joined the other interviewees in a waiting area that had the vibe of summer camp. The walls were orange too, covered with what looked like massive Lego blocks. We grabbed a free picnic table and scoped out the competition. Most of the other founders seemed around our age; some were furiously editing code on their laptops, while a few were testing out physical products. The room had the nervous feel of *American Idol*: of the dozens of applicants in this room, only an exceedingly small number would make it into the program, a fact everyone was well aware of.

When we were called in for our interview, we were disappointed to see that Paul Graham wasn't in our room. The questions streamed by so quickly that we hardly had time to blink. Ten minutes later, we left the room feeling entirely unsure of our chances.

But we were in California, and we had a car, so I suggested that we take Highway 1 south to Monterey, along the Pacific coast. The ethereally beautiful combination of sun and breeze and deep blue ocean under pale blue sky was marred only by the poor cell signal along the drive. My phone number was the one on the application, and if Y Combinator liked us, they would call to congratulate us; if they didn't, we would get an email. The others kept asking me if I had service.

Somewhere near the foothills of the Los Gatos mountains, I got a push notification. "I'm sorry to say..."

I told the others we got an email, and didn't even bother reading it out.

Later that night, after we'd all retreated into our hotel rooms, we each received a CC of an email from Nick with the subject "You guys messed up!". The email, sent to Paul Graham, asserted that our demographic inference engine was literally the best in the world at turning social data into insights, and that our business model of providing data to advertising platforms was equivalent to selling shovels during a gold rush. Over a stale continental breakfast the next morning, we made fun of Nick for his hot-headed bravado, but I was secretly impressed. If he cared that much, maybe we could actually pull this off.

By the time we got back to Montreal and its streets lined with slumbering mounds of brown snow, we decided that we didn't need Y Combinator's mark of validation after all. As long as we believed, we could figure out the rest.

FOUR: FAKE IT TILL YOU MAKE IT

As an entrepreneur, I try to push the limits. Pedal to the metal.
— Travis Kalanick in December 2014, while he was the CEO of Uber. He was ousted from his role several years later after a series of scandals.[1]

May: springtime, the snow melted, Montreal abruptly shifting from the bleak dregs of winter to blazing hot near-summer. I passed my last exam by the thinnest of margins and, after changing majors several times over the years, I would be graduating with a bachelor's degree in computer science and mathematics.

Most people in my year were planning to spend this summer travelling, hanging out, playing video games. I, however, was preparing to work harder than I ever had before, in search of something greater than myself. I would never admit this, of course; if you asked, I would give an evasive non-answer about wanting to work on something interesting with people I liked. But the truth was, I was lost and hoping that this last-ditch effort would save me.

Nick offered his apartment as our temporary office, and we painstakingly transported some second-hand desks and whiteboards up the creaking stairways. Nick's roommate was taking summer classes, and he had a tendency to glare at us over his cereal as we scribbled furiously on the whiteboard in his living room; Nick reassured us that his roommate was OK with the arrangement.

Our founding team mutated throughout the summer — our Y Combinator application included a graduate student in the lab who later decided he wasn't ready to drop out just yet. When the dust settled, our roles were decided: myself as Chief Technical Officer; Liam as Chief Operating Officer, which meant nothing but we wanted him to have a C-suite title; Tim as Chief Financial Officer, to start working later in the summer; and Nick as the prototypical nineteen-year-old college dropout CEO. The professor of the lab would be involved in an advisory capacity.

We weren't sure what our business model would be, but I didn't need to worry about that; my job was to build the underlying technical infrastructure. At first, I protested that I didn't really know what I was doing, but soon the joy of the challenge came rushing in. I sketched out infrastructure layout ideas in a grid-lined notebook, cobbling together insights from the different programming environments I'd worked in. To keep track of my progress, I started keeping a diary in which I laboriously recorded all my technical and non-technical achievements at the end of each day.

Figuring out our business model was Nick's job. He had never been a particularly enthusiastic programmer, but this was a task he was almost freakishly well-suited for. He had no qualms about asking people for advice or money, whereas I would rather die than send a cold email. He wasn't sure of our long-term vision yet, but he thought we could start by selling to all the social media analytics companies in the local area and iteratively learn from there, following the "Lean Startup" model proposed by entrepreneur Eric Ries. Nick was confident that we would be profitable by September.

Our first step was to incorporate. We set up as a Delaware C-Corp, which we knew was common for tech startups, as it simplified raising money from US investors and also lowered our tax bill. Our lawyers worked for a prestigious

Silicon Valley law firm used to handling startups, and they were happy to defer payment until we raised our first round. Incorporation was easy: a quick phone call, a few forms to fill out and scan.

The next step was a little harder. We wanted to move to the US, and particularly the Bay Area, because that seemed to be where all the cool startups were. But most of us did not have US citizenship, and a call with an immigration lawyer revealed that moving wouldn't be easy, especially if we didn't have money. Even after raising a round, obtaining visas could require employing legal trickery like pretending we had another office in Canada. While we were stuck here, we might as well focus on building a sustainable business. At least rent was cheap.

What's more, Quebec was apparently awash with government money for funding startups; every non-Quebec entrepreneur we met told us how envious they were.[2] Upon looking into it, we learned that we weren't eligible for funding unless we incorporated in Quebec. Nick liked the idea of setting up a Canadian subsidiary, but I was squeamish about it, protesting that this program was meant to stimulate actual science. As a startup founder, free money would be nice, but as a citizen, I wanted government funding to go to socially useful ventures. This startup might be a fun technical problem, but I was under no delusions about its social utility.

Midway through June of 2014, I came across an entrepreneur tackling exactly the kind of problem I wished we were tackling. *Fortune* had just published a glowing profile on Elizabeth Holmes, who founded medical technology startup Theranos after dropping out of Stanford at the age of nineteen.[3] I sent the piece to my co-founders with captions like "she's a boss right?" and "my hero". After ten years, her company had raised over $400m, swelling to a valuation of $9 billion. Her former Stanford professor was on the board, as well as someone named Henry Kissinger who sounded important. The parallels to my startup — our CEO also being a nineteen-year-old college dropout, starting a company with his professor — were not lost on me, even if our technology was much less revolutionary, and our university less prestigious.

On the other hand, it took Elizabeth Holmes a decade to get to the point where she was getting glowing write-ups in prominent magazines. We were less than two months into our startup and I was already getting fed up. In June, Nick went to New York to meet with some advertising technology company he had promised would be our first customer. The company wanted a demo first, and Nick called me in the middle of the meeting on Monday morning to convey their request.

"They want these users classified ASAP", he explained to me over background static.

"But I told you I'm in the middle of a rewrite", I responded, incredulous. "I can't do that right now."

"Drop everything you're working on and get this done", he insisted. "I think we can close this."

I hung up, seething that Nick had once again promised something we didn't have. There was a certain logic to it: if they didn't bite, I wouldn't have wasted time building a useless product, and if they did, we would just work really hard to get it done in time. Still, Nick's fake-it-till-you-make-it, ask-for-forgiveness-not-permission attitude was starting to get on my nerves, especially when I was the one who had to do the actual work whenever we sold something that didn't exist.

In the end, I couldn't get the system to work in time, so we had to manually generate their data. It was annoying, messy, and possibly unethical, but no one would ever know; it was only sales, after all. But I wasn't happy about it, and I was even angrier when I found out that the sale was more uncertain than Nick had made it seem. Later that day, I sent Nick a flurry of irate Slack messages explaining that I couldn't work with him if I couldn't trust him. Nick responded that he was just trying to do what was best for our company; he worried that we were moving too slowly. That was fair. I resolved to give it another shot.

In July, we bonded over a popular comic graphing the ups and downs of the startup journey, depicting a trough of sorrow that would eventually give way to the promised land where everything would go up and to the right.[4] We were sure that we would get there eventually. At the moment, though, we were still getting used to each other's working styles and habits, and it was tough. The atmosphere in the room swung wildly between fever-pitch enthusiasm and dark, monotonous lows. We still didn't have a functioning product, and our only contracts were of the noncommittal sort; all we had in our bank account was a few paltry thousand that we had won in a university-wide entrepreneurship competition. In this uncertain atmosphere, the tiniest thing could tip us over into the trough.

One day I came across a *WIRED* profile of some startup founders living in a hacker house in San Francisco.[5] They were young, smart, hardworking; they cycled between optimistic highs and cynical lows; their startup didn't make it. The piece detailed the founders' gradual lowering of their aims, from their initial sky-high ambition to build something world-changing into finding themselves creating a banal e-commerce plugin based on the advice of their investors. Not only did they not become millionaires, but the autonomy they thought they had as founders was only illusory, a shroud to conceal the truth of a system primarily benefiting those who already had money.

Reading this piece hit me like a brick, and I suddenly didn't feel motivated to get back to coding. Rationally,

I knew we were almost certainly going to fail, but the only way to motivate myself on a day-to-day basis was by deliberately behaving irrationally — by pretending that we were one of the small number of startups that would succeed, and pouring my heart into it accordingly. I wondered how much of the new tech boom was propped up by the naive dreams of aspiring startup founders like us, living off savings or credit card debt, working on problems they didn't *really* believe in because they thought it was the only way to succeed — because they saw this as the path to fame and fortune, or at least a less mundane alternative to working a 9-5. When really, they were just doing cheap R&D for other companies.

I gave myself a few minutes to wallow in the trough, and then I pulled myself together so I could go back to the new testing framework I was prototyping. I could indulge in melancholy after we'd exited. Right now, my to-do list awaited.

By July, despite the occasional hiccup, things seemed to be going reasonably well. We were further behind than we had hoped, but at least we liked each other enough to want to keep working together. Tim hadn't started working yet, so Liam, Nick and I planned a trip to San Francisco to meet with potential partners and hang out with other founders. It was also an excuse to soak up Silicon Valley culture, because Montreal was light on tech startups and we would rather be where the action was.

Our first meeting was with a mobile advertising company in SoMa. We over-gorged on chicken and waffles at a nearby dive bar beforehand, triggering a queasiness that didn't go away for the rest of the day. Their office was a typical startup loft, with a faux-industrial warehouse aesthetic, open-plan floors, and conference rooms with quirky names. We were supposed to meet with the "VP of Talent" (this is really how startup titles work) and we weren't sure if he wanted to invest in us, partner with us, or acquire us. After a few minutes of the meeting, we all independently decided we wanted none of these. Said VP didn't seem like he had much interest in discussing our relationship, either — most of the meeting featured him laboriously detailing his company's brilliant business model, which was generating his employer unbelievable mountains of money through facilitating full-screen video advertising on mobile. He seemed almost astonished about it all. It was as if he had to keep saying it out loud in order to absorb it.

After an hour of this unwanted monologue, we stumbled out of there in an odd daze, pausing outside the door to adjust our bearings. It was sunny, and the sky was a high, cold blue, and the air had that weird San Francisco summer chill that always came as a surprise. We looked at each other: what the hell just happened? Why did we subject ourselves to an hour of nodding along to a self-important startup dude's interminable boasting about his company's financial prowess? Was this what startup life would be like? Talking to each other helped dissolve the stress; our path might be hard, but hey, at least we weren't working for *that* guy.

In the evening, we headed for a party in Bernal Heights. The person who invited us was a young founder whom Nick had recently connected with on LinkedIn, and we didn't know what to expect. It turned out to be a hacker house inhabited by about a dozen people in the startup scene, all founders or soon-to-be founders waiting for the right bolt of inspiration to strike. One of the inhabitants was a Thiel fellow — meaning that PayPal co-founder and controversial billionaire Peter Thiel paid him $100,000 to put off university — and he gave Nick tips on his own Thiel application. I was the only woman present, and besides us, there was only one non-white attendee, who didn't live in the house; as the night went on, his role began to feel like that of a punching bag for the house's inhabitants.

Someone declared an impromptu pitching contest, and a projector appeared out of nowhere. The other non-white guy was picked to go first. He had recently moved here from East Asia, and his English was pretty elementary. Nobody seemed to fully understand what his startup did, and he wasn't fluent enough in the cultural norms to answer the audience's questions in a way they found satisfactory. Soon the vibe turned cruel, and there was an unrestrained glee in the audience when the pitcher stumbled over words or missed the subtext of a question. Nick was drunk and didn't seem to notice, but I was even more uncomfortable than when I first clocked the party's gender ratio, and Liam didn't look happy, either.

Finally, someone cut the humiliating spectacle short, and it was Nick's turn. I was worried that he would be eviscerated too, but Nick's English was fine and the audience could tell he was one of them. It was less what he said and more the way he said it: his speech was sprinkled with casual expletives, and he clearly wasn't trying to impress them, which was what impressed them.

Pitch over, I called us a Lyft and nudged Nick toward the door, citing early morning meetings. As Nick snored in the front seat, Liam quietly declared that he didn't want to move to San Francisco if these were the kind of people we would be hanging out with. I shared his unease, but I wasn't quite ready to agree. Maybe hacker houses just attracted the worst bits of San Francisco? I still wanted to move here, I told him; after all, what other place would be as appropriate for the path we'd chosen?

The midnight drive through the foggy hills was unexpectedly peaceful. Our Lyft driver silently steered us through the city, occasionally glancing at his phone for directions. In the distance, flashing lights danced up and down the length of a misty bridge.

Back in Montreal, things were ramping up. Our work environment was squalid, but that was almost glamorous in itself; I was certain that the founders I idolised had started out just like us, working out of a dishevelled living room surrounded by coffee cups and take-out containers. I frequently stayed at the apartment past midnight, occasionally taking a break from coding to read startup books on the futon. Even a recurring ant infestation didn't ruin the vibe — sometimes I just had to travel around the apartment by dragging my swivel chair while Nick doggedly poured laundry detergent over the cracks in the floorboards.

One night, it was just Nick and I in the office. As the clock ticked closer to midnight in a room littered with ant carcasses and plastic wrappers, Nick explained that we would all have to change quite a bit if we wanted to build a successful company. I told him I agreed, though in all honesty I wasn't sure if I was ready for that. He responded that we would have to if we didn't want to get stuck running a lifestyle business which capped out at a measly $10m in revenue. I laughed, reminding him that it was silly to denigrate the idea of a company making millions of dollars when we had precisely $0 in revenue, but I knew he was right. If we ever did become a mid-tier company, with only modest ambitions, we might as well sell it off, move onto something bigger and better.

By August, I had decided: I was ready to do this. I knew this might just be another wiggle of false hope in the startup lifecycle, but I really thought we could make it. I had never grown so much or had so much fun in such a short period of time. Even though my savings were being depleted at an alarming rate, it still felt 100% worth it, and I wouldn't have traded it for anything.

I felt bad about the prospect of reneging on my Google offer, but it was an at-will employment contract; reneging was only a little worse than quitting on my first day, which would be valid, if bizarre. My recruiter was surprisingly understanding about my change of heart, telling me that the door would always be open if I changed my mind. That was a relief; I had worried that I would be blacklisted. It almost felt unfair that I could get away with so much.

In the meantime, we were developing a clearer understanding of the industry and our potential place within it. One team meeting ended with Nick drawing a diagram on the whiteboard showing other companies in the space as verticals, and a wide box underneath. The box could be us, Nick explained excitedly; we could be the horizontal layer underneath everything else, the data platform on which all these other companies depended. We were months out of school, and we didn't even have a functioning product, but already we could see a path to becoming a billion-dollar company through charging rent to other companies that needed our data. Why not? Other founders had done the same thing to other industries; weren't we just as good as them?

(Never did we ask ourselves if our vision of dominance over the ecosystem was good for anyone else. Societal good was never our framework. All we cared about was how *we* could dominate, with our technology being a means to the end of extracting rent from every transaction.)

The summer sped by. Bulging garbage bags piled up in the kitchen and were periodically removed to reveal entire ant colonies underneath. We started communally watching episodes of HBO's *Silicon Valley*, and even though I knew it was meant to be making fun of Silicon Valley culture, I still found it unexpectedly inspirational. We were working nights and sometimes weekends, convinced that success was just around the corner — as soon as we pushed this new feature or closed this promising lead. I believed this even when I was spending half my time putting out fires as we continually strained the limits of our system, rather than actually making progress on building new features. But surely that was normal; every successful startup had stories of the servers being on fire half the time. Hell, Facebook's motto was, until recently, "Move fast and break things".[6]

Many days, I witnessed the office clock hitting twelve twice. The short walk home through the quiet tree-lined streets was restorative; I felt drained, but in a good way, like I had set a personal record. We still didn't know what we were doing, but I was a staunch believer in the transformative power of hard work, and I was sure we would figure it out by the end of this redemptive baptism by fire.

But my doubts were never completely quelled, and one night I talked to Nick about them. Could we really do this? We had only just graduated from university — dropped out, in Nick's case — and yet we thought we could make a dent in this huge market. I personally needed to build a robust inference system that outperformed competitors with many years on us, way more personnel, millions more in funding. In our post-Zuckerberg era, I knew that it was theoretically *possible*, but I was still afraid to drink the Kool-Aid. It felt delusional to believe that we could actually succeed.

But we *could*, Nick insisted. The companies that currently dominated weren't invulnerable; if we worked harder than them, if we improved faster than they could improve, if we were leaner and more willing to take risks than they were, then we could win.

But big companies had internal guidelines, professional

standards, HR to turn to if things went wrong; our team was tiny, with haphazard policies and no accountability mechanisms beyond ourselves. On the bleakest days, when Nick had messed up scheduling another meeting or broken another spreadsheet, I was on the verge of giving up. All we had was our trust in each other, and sometimes that frayed. And when we were all mired in misery, it was easy to lash out.

One day, Nick made a minor mistake that pushed me over the edge, and I snapped: "You know I turned down Google for this, right?" That was my trump card, but as soon as I said it, I felt a stab of guilt. I knew Nick was doing his best, and anyway it wasn't really true. I didn't actually turn down Google for this. This startup was just the excuse; I probably would have jumped on any half-decent excuse that presented itself if I had waited long enough.

Why had I really turned down Google? I was still piecing together the answer myself. Part of it was that I could not find a way to reconcile my ideological commitment to open source with working at Google; another part of it was a vague discomfort over the inequality I had seen that summer in San Francisco, and which I hadn't wanted to actively contribute to. But underpinning it all was a nameless dread of imprisonment, which I knew was irrational and conceited and yet could not seem to shake. In the end, I couldn't bear the thought of going back.

But this startup — my startup — would be different. This startup would be exactly what I needed.

September drifted along in a rush of auburn leaves and cool nights. We were supposed to be profitable by now, but we still didn't have a single paying customer.

When things weren't going well, it became harder to maintain team cohesion. Disillusioned, the sugar rush wearing off, we sniped at each other more frequently. Liam had been working on a new clustering feature in the hopes of winning over a potential customer, but one day Nick declared that Liam's work was useless, and that he should instead be working on a mini-pivot to advertising technology that Nick had unilaterally decided we needed to commence. I was getting sick of Nick's erratic management, but I was embarrassed by how little I knew about pixel-tracking, and anyway I feared that he might be right, so I settled for writing a recap in my diary drenched in passive-aggressive hostility.

The next day, Nick told me I was moving too slowly when I should have been getting something out the door ASAP. I was immediately aggrieved, but I couldn't really disagree. I responded that I understood the business imperatives, but I still wanted our tech to be good to separate us from competitors. That wasn't what truly motivated me — I wanted to build good tech for the sake of good tech, irrespective of business value — but I knew better than to explain that to Nick. I *knew* that business metrics were the only ones that mattered in the end; I just hated that this was true.

In the meantime, we kept up with news of successful startups like they were dispatches from a faraway island we wanted to visit. Liam and Nick especially idolised SpaceX, and whenever there was a particularly exciting SpaceX rocket launch, we would take a break and gather around Liam's laptop to watch the live stream. So after hearing about Elon Musk's ambitions to build a colony on Mars, Nick told us that this was one opportunity he would consider ditching our startup for.

"What, live on Mars?" asked Liam doubtfully. "You know you can never come back, right?"

"That's fine", responded Nick cheerfully. "I'd work remotely."

And then, miraculously, a switch flipped and things started picking up. A major cloud hosting company offered us free hosting for a year, as part of a program for supporting startups; we could finally move off the tiny server we'd been using and spring for better hardware. And someone whose company recently got acquired by Salesforce — another one of Nick's successful LinkedIn connections — stopped by the office one evening to tell us we should try to get a contract with them too, as they were in the process of negotiating with one of our competitors. That would be punching way above our weight, and it would be a major commitment, but it would mean money and a great customer name for our pitch deck. We weighed it up with all the gravity of a decision between equals. The possibility that our technology might be good enough to get us a contract with a company as successful as Salesforce was enough to raise our spirits.

School had started again, and it felt strange to be surrounded by first-years blasting dubstep and shotgunning cans of Molson while we were trying to work. Slowly, the technical infrastructure got refined, and my data inference engine got better and better. Our Python codebase was infesting my dreams; one night I was attacked by a 140-foot python. A different dream featured me trying to fix bugs in a feature I'd been working on, my focus narrowed to a single elegant line of code; when I awakened, I got out my laptop to confirm that my dream solution worked in real life. And one day, I tested my classifier on some real-world data only to find that it had inferred some attributes better than I could have inferred manually. I excitedly told the others that our system was starting to attain sentience, though maybe that was just projected narcissism.

As long as the system was improving itself, I could feel good about myself, too. I wouldn't have to feel guilty about taking a night off, because something tangentially associated with me was improving. If I wasn't improving, then I would be standing still, and that was the worst thing I could imagine. Through this system, though, I could achieve a kind of immortality, allowing me to transcend the petty realm of business matters like product-market fit or raising money.

In November, we visited San Francisco again. We liked finding excuses to come to San Francisco, with its balmy weather and ambient hum of startup success. This time, the excuse was *not* getting an interview with Y Combinator — we'd applied again, even getting alumni feedback on our answers, to no avail. But we decided to go to California anyway, giving the trip the moniker of Y Not. We were outwardly blasé about the rejection but really it hurt; we needed guidance, because it felt like we were spinning aimlessly.

Nick had arranged a bunch of meetings for us on the day after we arrived. The first meeting was in Palo Alto, and in the morning we took the Caltrain down just in time for lunch on a rooftop terrace under the soothing California sun. The meeting was with a founder of a Y Combinator-backed startup building tools to optimise social media marketing. Their business was about as exciting as ours, and their elevator pitch sounded every bit as enthusiastic too — which was to say, not very, though we all put up a good front in feigning it. Still, it was nice to spend some time with another young founder in our space, especially one whose business was further along the path to success, and we drank up his advice as if it were a tall glass of water in a desert. It wouldn't get us out of here, but at least it might keep us alive long enough to find a way out.

Our next meeting was in San Francisco proper, with a marketing startup that was bigger, hotter, better-funded.

Their office had floor-to-ceiling windows with jaw-dropping views of the city, and for a second I wondered if they were hiring. We had met with them before — last time there were four of them meeting with two of us, and I had been simultaneously flattered and alarmed that they thought we were worth devoting four employee-hours to. This time, there was only one of them, and the meeting started off on the wrong foot as it turned out they had misunderstood what our product did. No problem, that happened all the time; sometimes we weren't sure what our product did, either.

Things brightened up when the executive mentioned that they were using one of our direct competitors. That was our cue to eagerly launch into our spiel, throwing out accuracy numbers from various comparison tests and citing testimonials from customers we'd convinced to switch away. We left the meeting feeling hopeful, though not outright elated. Even if we closed this client, it would only be a small contract, and we needed a rocket ship if all this was going to be worth it.

Our next meeting was on the other side of Market Street. It was a deceptively old building; taking the elevator to the top floor revealed a typical startup office, which you would never have guessed from the outside. We were here to see a founder around our age, whose company was way ahead of ours — they had an office, money in the bank, the Y Combinator badge of honour. The founder's emails to us were filled with phrases like "vision" and "trajectory" and "big play", and he was about as bombastic in person, telling us how proud he was to have an idea that he deeply cared about, and why didn't we have an idea we cared about as deeply? His company helped brands understand the ROI of their advertising campaigns. He had the slightly glazed look of someone who had lucked into tons of money and, deep down, was worried that he would have to give it back.

The lighting in the conference room was sparse, and staying awake during the meeting soon became a challenge. It wasn't clear why we were here, if this guy was just going to keep bragging about his brilliant idea and his many millions in the bank in lieu of actually having a conversation. Hadn't he said in his emails that he wanted to integrate our product with his? He did use the term "joining forces" multiple times in the meeting, but whether he meant in the sense of a product integration, or a talent acquisition, we weren't sure. When the meeting was finally over, we debriefed and collectively decided against following up.

It was 6pm, and we still had one meeting left. We were a little drained, and, given how the last meeting went, not feeling especially optimistic. But this was a social visit to a cosy live/work loft in SoMa to meet the founders of a small Y Combinator-backed startup. We grabbed beers and walked up to the roof, soaking up the 360-degree view of the city. Maybe it was the view, maybe it was the beer, but this was the best conversation we'd had all day. These guys had a similar startup to ours — also marketing tech, also kind of boring — but for the first time all day we didn't talk about work, and there was a real sense of kinship that I hadn't felt in a while. Not for the first time, nor for the last, I was filled with hope that we might make it after all.

Fall turned into winter. Red leaves cascaded off mottled branches and slowly disintegrated into the sidewalk. Temperatures plummeted until the air bit like pinpricks. Liam, Tim and I had recently become roommates, and we lobbied to move the office to our living room, instead of Nick's. It made sense from a utilitarian perspective: one person walking through the snow every day was fewer than three. Plus, we'd never seen any ants in our place.

At other companies in our space, winter was usually

a sleepy time of the year due to the impending holidays, but we were determined to work harder than ever before. One weekend, Liam and I decided to throw our own mini-hackathon to blaze through our to-do lists, and we stayed up working until 4am. After that, 4am ceased to feel like an unusual time to be awake, and the lines under my eyes started to settle in. One day I picked up a particularly nasty cold, but I still forced myself to grit my teeth and work through the fog. Suffering would only make me stronger.

Nick's job was to get us meetings with other companies. We would usually begin the process by feeling either positive or neutral about the company, but after a couple of meetings, our attitude would usually turn into animosity. We needed their money for validation and growth, but they were always stringent with it, and we hated that we had to depend on their largesse. In private, we would insult their product design and mock the misspellings on their website; in meetings, we would meekly answer questions and agree to their demands. I knew it wasn't a great sign that our primary attitude towards our customers was contempt, but whatever, we would make it up in volume.

That pretence of being dutiful worked, in a sense: several of the companies we were working with said they admired what we had achieved and wanted to join forces. On the one hand, that was a self-esteem boost; on the other, we resented that they believed they could acquire us — as if we'd be content working as a mere R&D lab for someone else's lifestyle company. Didn't they realise that our ambitions far exceeded theirs?

My pride was mixed up in this. If I was going to be working for someone else's company, I might as well work for Google over a small Canadian firm that no one had ever heard of. There was a hierarchy in my mind, and the shining hill of Silicon Valley, with its free-flowing cash spigot and venerated brands, sat firmly on top. Part of me suspected that this was an unhealthy way to think, but I didn't know how to view the industry except through a financial lens.

We tried not to get too distracted by acquisition offers, as we had enough on our plate merely deciding on a product. Every week Nick would come up with a new niche for us to explore: one week it was a suite of unspecified services to improve the Twitter advertising experience; another, it was augmenting social share tools; another, it was helping consumer brands understand their customer base using clustering. By January, Nick was convinced he had found our niche, for real this time: e-commerce. My job was to do market research and figure out how to integrate our technology, and though I had zero interest in the space, Nick harangued me into taking a break from coding to conduct a desultory investigation of companies with names like SailThru and TellApart.

Liam, on the other hand, hated e-commerce with a passion. Instead, he thought we should make tools to help content creators better understand social media. The #gamergate campaign was in full swing, and we'd followed it avidly, less as a cultural phenomenon and more as an occasion to test our technology. Forget the potential implications of an army of disaffected men building a movement around harassing women — we just wanted to capitalise on it with a viral product.

Nick wasn't as bullish on the idea of selling to individuals; he had a bee in his bonnet about enterprise sales. So Liam suggested a compromise: an audience discovery tool that he described as "Tinder for advertisers". We pitched this to an advertising agency coincidentally located just around the corner, and they asked if we could use this to differentiate between the "Vanessas" and the "Monicas" in a brand's customer base — the former, professional working women who were happy to splurge on designer clothing; the latter, mothers more interested in home goods. My next few days were spent looking through the Twitter followers of accounts like *Vogue* and Sephora to find candidates for these cartoonishly vague marketing personas.

Was this constant pivoting a sign of being outmanoeuvred by bigger companies wanting to outsource grunt work to a naive startup? Or was it a principled, "Lean Startup"-approved strategy for inching towards product-market fit? What was the difference? I mostly ignored the business side to focus on the battleground of our technical infrastructure; I was up to my ears in customer complaints about our API. The others could figure out our go-to-market strategy.

February of 2015 felt like the seasonal equivalent of the trough of sorrow: the bitter wind howling at the horizon, the icy air making every breath sharp. As the weather worsened, so did our mood. Nothing was happening and no one was sure of anything, and in the absence of external distractions we took out our anger on each other. I snapped at Liam for bungling an implementation of neural networks; Liam was mad at Nick for continually flip-flopping from idea to idea; Nick was frustrated with the rest of us because we weren't pivoting quickly enough. I thought Nick's behaviour was borderline unethical, possibly the result of having read too many Steve Jobs biographies; Nick thought he was doing what was necessary to motivate us. The mood blackened, our bickering sometimes disintegrating into a teamwide silent treatment.

One week, when the atmosphere was particularly sour, it happened that our main server encountered severe hardware failures. There was nothing we could do about it, and there was nothing we could do without it, so Liam and I decided to dull the pain by gaming. We returned to an old MMORPG we had both independently played as kids, and spent several days submerged in the idyllic world of RuneScape, XP-grinding by clicking on blocky pixels for the small joys of achieving quantifiable progress towards a defined goal. Even after the hardware issues were fixed and we could return to work, for a few weeks we kept a browser window open in the background in order to do some passive levelling-up while working. Nick occasionally looked at us askance, but he never said anything; it had been a long winter, and we were all flagging.

Spring again. The snow melted and it was immediately searing hot. In this fresher air, I felt cautiously optimistic again. In April we got our first paying customer: a political consultancy startup with ties to Barack Obama's 2008 presidential campaign. I actually liked this company, and it was satisfying to have their vote of confidence.

Still, our total revenue was peanuts compared to what our competitors were raking in. One afternoon we gathered around Nick's computer to pore over the website of one particularly hot competitor, which looked concerningly similar to the way we marketed ourselves. And maybe it was paranoia, but we saw future competitors everywhere — every consumer app that shot to the top of *Product Hunt* could potentially become a rival, if they ever decided to sell the data their users had unwittingly provided them. Nick thought we should get ahead of this by building our own viral app for the purpose of secretly collecting user data, but we didn't know what we would build.

The competition scared me — it felt like everyone was speeding by while we were standing still. And the sheer amount of competition had me asking some scary questions, like why was I doing this in the first place? Why was I building tech that other people had already built? Questions I'd rather not think about, when I'd already sunk a whole year into it. In my to-do lists, sandwiched between outstanding bugs to fix and features to build, were desperate cries for help like *Figure out the direction of the company* and *Decide on personal plans if this doesn't work*. We weren't drowning, exactly, but we weren't going to be able to tread water for much longer.

A biography of Elon Musk was coming out; I placed a pre-order on Amazon. Even if the book didn't have advice specific to my plight, at least it might give me motivation, and I needed some of that.

FIVE: ACCELERATE

Work like hell. I mean you just have to put in 80 to 100 hour weeks every week. [...] If other people are putting in 40 hour workweeks and you're putting in 100 hour workweeks, then even if you're doing the same thing, you know that you will achieve in four months what it takes them a year to achieve.
— Elon Musk, co-founder of PayPal and now CEO of both SpaceX and Tesla, when asked to give advice for entrepreneurs in a December 2010 interview[1]

The first anniversary of our startup's incorporation loomed, then passed. We were still drifting aimlessly, the timeline for profitability slipping back further and further.

When we started, we thought we would be cash-flow positive within four months. But a whole year later, we barely had a saleable product, and our revenue to date was a grand total of $125 USD. And yet we were consumed by a high fire of certainty that the mythical product-market fit was just around the corner; if we kept grinding away, at some point it would all make sense.

So it felt like a bolt of opportune lightning when we heard about a startup accelerator that was looking for another startup. Their next batch was due to start in a few weeks, and one company had to drop out last minute — did we want to replace them? Hell yes we did. This accelerator wasn't as prestigious as Y Combinator, and they were based in New York City rather than the Bay Area, but their portfolio looked decent, and the partners asked thoughtful questions over the phone. Maybe this would be our saviour — structure, legitimacy, advice. People with money and experience and wisdom in our corner, literally invested in our success.

Everything else was starting to line up, too. On the cusp of the accelerator, our cloud hosting vendor extended our free hosting by another year and even threw in a complimentary machine upgrade. Several potential customers were inching closer to long-term contracts. And as a team, we were closer than ever, the toxic atmosphere of last winter having melted away like snow under the blazing sun.

Our main hurdle was legal. I was a stickler for legality, and my fear of conducting an immigration misstep bordered on paranoia. A quick Google search brought up the story of a Canadian founder whose acceptance into Y Combinator was curtailed when he was denied entry to the US. This triggered a deep dive into researching whether attending a startup accelerator counted as "working illegally" in the US, but all I concluded was that it was an extremely grey area, and very open to interpretation; immigration law wasn't exactly up to date with trends in the startup world. Anxious at the possibility of being denied entry, I fastidiously printed out official documents proving that I was allowed to be in the country for a business incubator, ready for an onslaught of questions. Nick, confident that everything would be fine, laughed at my overpreparation.

In the end, we were both wrong. At the airport, I got a cursory ten seconds of questioning by a border agent before being dismissively waved through, my thick sheaf of printouts still neatly folded in my carry-on. Nick, who had opted for a land route, got pulled off the Amtrak train by sceptical border agents; they questioned him for hours in an airless room while thumbing through his phone. By the time they let him out, the train had long departed, and he had to hitch a ride from the border checkpoint. At least we all got through.

The next immediate task was to find a place to live. The accelerator office was in Midtown Manhattan, and we wanted to live close by, but Manhattan was not exactly cheap, nor was it easy for outsiders to find months-long lodging on short notice. We booked an apartment in Murray Hill via Airbnb, sight unseen, and when we arrived we determined that this so-called two-bedroom apartment had originally been a one-bedroom, with the extra bedroom created by walling off a portion of the living room. The remainder of the living room was consequently devoid of natural light and contained a worn mid-century couch, a flimsy coffee table, and an ageing wooden chair. The kitchen had approximately one square foot of counter space and the small refrigerator was crammed into the coat closet. We paid $5,246 USD a month.

I was oddly happy about our lacklustre accommodation. Maybe it was a blessing in disguise — the less we enjoyed being at home, the more natural it would become to spend time at work, which was what I thought we should prioritise. We weren't here to *enjoy* ourselves, after all; we were here to work in order to make something of ourselves. Enjoyment was for people who wanted to fail.

The rest of June passed like a fever dream. We would wake up to horns blaring from the frustrated drivers on 1st Avenue and take turns brushing teeth under the suspiciously organic-looking material in the bathroom skylight before leaving together to make the thirty-minute walk to the office. The towering skyscrapers that lined our daily commute felt momentous. The stakes were bigger now that we were in an office humming with other founders, distinguished visitors periodically dropping in to give advice. It certainly made one less inclined to play RuneScape.

As great as it was to be around other founders, it also made me anxious. The median age in this batch was somewhere around thirty, and most of the others had work experience, connections, business degrees. We, however, were the youngest in the batch, with neither recurring revenue nor evidence of product-market fit, and we were living in a 828-sq ft apartment that we suspected was harbouring a mould infestation. All we really had was a burning conviction that we would succeed, coupled with the willingness to do impossibly stupid things to prove it.

Our business model was still stuck in the product equivalent of throwing spaghetti at the wall, guided by whichever latest marketing trend had caught Nick's attention. By the end of June, Nick was convinced that he had found our first big customer: a content creator company based in Los Angeles. I peered uncertainly at their website, heart sinking with every stock photo and meaningless marketing buzzword, realising that I had no interest in working for this company, or really, any of the companies we'd been talking to so far. But it wasn't like I had better ideas.

One morning, Nick told us that we could get a contract with this content creator company if we agreed to build a minimum viable product in two weeks. He was so excited about the prospect that I felt bad having to tell him that we couldn't, because we were in the middle of an infrastructure overhaul: our system was not made to handle our current usage levels, and with each day we were inching dangerously closer to overwhelming it. Liam and I were spending all our energy on setting up a parallel system using a distributed database. It didn't feel like progress, but we had been living on borrowed time anyway.

July stormed in, hot and humid. The office AC was running full blast: the accelerator's partners, all male, liked to wear the same starchy shirts and jeans that they wore year-round; the women, more likely to adopt summery linens and other weather-appropriate attire, soon learned to bring cardigans. We'd all gotten used to the indignities of working in a co-working space surrounded by other founders taking calls and occasionally playing ping-pong.

The accelerator frequently hosted events for the New York startup community. Attendance was not compulsory, but most of us went anyway. One evening, a panel of accelerator alumni shared their success stories; one panellist, a woman in thick-rimmed glasses whose startup sounded kind of similar to ours, spoke confidently about machine learning and data-driven decision-making. That could be me next year, I thought, up on that stage, fresh off a new round of funding, secure in the knowledge that we were on our way to the promised land.

In the meantime, we hadn't even decided on a product. Everyone had their own pet project, and it was a bit like being on a boat where everyone was paddling in a different direction; tensions were accordingly high. Liam thought we should focus exclusively on our API; Tim wanted to work with advertising agencies; Nick was certain that we would become a billion-dollar company if we simply built the missing piece in the ad-tech ecosystem. Meanwhile, I was engrossed in poring over the database and didn't really care what we sold, to Nick's clear exasperation. I knew we needed to find product-market fit, I told Nick, but unless I fixed the database ASAP, all the work we had done so far would go up in smoke. Nick was frustrated with me for ignoring the business side and I was frustrated with the technical limitations, the uninspiring business prospects, and myself for being in a situation that I didn't like but couldn't see a way out of.

But then Nick found a lead in a more promising direction, the fruits of a quasi-superstitious cold email strategy that included sending follow-ups late at night to ensure they were at the top of the inbox in the morning. Unlike the marketing companies we had been working with recently, this company felt more like a traditional tech company. They were well-known in our space, and I was flattered they were deigning to talk to us. They asked us how much we would charge to include our data in their offering, which should have been a simple question, but it sent us into a frenzy of debate. We camped out in a conference room and threw numbers onto a whiteboard: $5,000? $15,000? $20,000? After a few minutes, we were convinced that our value proposition was good enough to support at least $20,000 a month.

A partner from the accelerator program poked his head in to give us advice that swelled our egos even more. If we were going to be adding more than $20,000 worth of value, he said, why would we be content to sell our data for only $20,000? We started thinking about alternative business models, like a revenue share, which would be more work in the short term but might make us a good target for an eventual acquisition. The company seemed confused when Nick came back with our proposal: they weren't used to potential vendors pushing for a 50/50 partnership, especially when that vendor was a barely-funded startup with no employees. But Nick was relentless in a way that seemed to impress them, and they agreed.

That partnership wouldn't materialise for a while, but in the interim we could put their logo in our pitch deck. Like most accelerators, this three-month program ended with a Demo Day, where each startup in the batch would give a five-minute pitch to potential customers and investors in the audience. Nick, as CEO, would be giving ours, but we couldn't really prepare a pitch without a solid business model, which meant our pitch was still incoherent. That had been acceptable in June, but by July, the other founders had refined their messaging, and our unpolished deck stood out unpleasantly during the weekly pitch practice sessions.

The biggest problem was that our pitch was boring. The partners looked bored, the other founders looked bored, *I* was bored and I was the one who was supposed to be building this vision. Nick rambled about fixing marketing while speeding through a Keynote presentation with cartoonish graphics and perplexing numbers. He hadn't shown us the deck beforehand, and we were aghast at how poorly he was representing us. The recriminatory team meeting afterwards devolved into a blame game where all our mounting issues spilled out: we told Nick he had to stop developing plans in secrecy; he told us that we practically forced him to by not being receptive to new ideas.

It felt like we kept having these talks. The constant pivoting, the atrocious living situation — it was beginning to wear on us. But there was nowhere to retreat to, because we were all committed to this; hell, we all lived together. We ended the meeting with an uneasy detente where we each agreed to compromise.

It worked: next week's pitch was much stronger, with a coherent message about our expertise in mining social data to get more personalised marketing. At least, that was what the pitch deck said; it wasn't clear to me how much the pitch deck was meant to reflect reality, versus simply a glossy version of ourselves that we sold to outsiders. Secretly, I thought it was kind of BS, and I worried that Nick was spending so much time working on the deck that he had come to believe it. We all knew that this "smarter marketing" stuff signified exactly null, right? That was just what we said to get people to invest, surely? The truth was, I didn't know what else we were other than the exaggerated claims and rounded-up numbers of our uninspiring pitch deck.

I was probably overthinking it; none of our competitors seemed to have a problem with the mundanity of our space. Word on the street was that one competitor was a rocket ship, and that they were eyeing the sales team of another competitor. We didn't have a sales team but maybe they would buy us too. Nick had somehow snagged an invite to a party they were hosting in the city, and afterward he drunkenly regaled us with assurances that everyone who mattered in the industry knew about us. "We have a safety net now," he told us excitedly. "Even if we fuck up, one of these guys will still acquire us."

How much would we sell for? We exchanged semi-serious numbers: a million? Twenty million? Fifty million? Tim didn't think it made sense to sell yet — he thought we didn't have enough information to know what we were worth. Nick proclaimed that if we were to get an acquisition offer for fifty million, he would personally tattoo a phrase of our choice on his forearm. Liam then put forward a counter-proposal of forehead, which Nick said he would have to think about.

It was unlikely, of course, but every successful acquisition had been unlikely at *some point*. Over the last few months, we had attended so many talks featuring founders who had lucked into wildly successful acquisitions. It was probably not going to happen to us, sure, but it *might*.

And I would never have said this out loud, because I knew how arrogant it would sound, but hovering in my mind was the conclusion that I had been right all along. About this startup, about the industry, about the untold hours spent alone in front of my computer, making silly websites and reading Paul Graham essays. People like me were accruing money, and power, and social status — they were essentially being rewarded *for being nerds*. I'd finally alighted on the right path.

By August, we fell into a rhythm of overtime work that must have looked absurd even to the other founders. The office was usually empty on weekends, and outside the golden summer days stretched for eternity, but we would be inside, hunched over our desks, sacrificing our youth at the altar of startup success. My days started to blend together in a hazy swirl of frenzied coding sessions and startup networking events, with the neon-lit streets of New York City merely serving as the unimportant backdrop for our all-consuming quest.

I was utterly entranced by thoughts of our inference system, my spare CPU cycles devoted to fixing bugs and architecting new features. When I looked at people, I would see the demographic and psychographic attributes we could infer about them. The real world was big and messy, but my little world of support vector machines and well-formatted commit messages was neat and orderly.

One day, we decided to expand to other social networks beyond Twitter. (We had assured our partners that going cross-platform would reduce platform risk, though at best it was like buying one house in an earthquake zone and then another in tornado country.) The problem was that most other networks had stringent data policies, making it difficult to get their data. Instagram had seemed like the best bet, but reading their terms of service, I discovered that while they might technically allow us to store user data, they prohibited the use of this data for targeted advertising. Pulling up Twitter's terms of service revealed nearly the exact same clause, which we had somehow missed. Both platforms expressly forbade us from doing what we'd been pitching for the last few months.

I shared the grim news with the team, and for the next several days we were plunged into a pit of despair. We looked for loopholes: *programmatic* advertising might be prohibited, but maybe we could get away with other forms of marketing, like email? But the wording was broad enough that any form of marketing based on this data could be considered against their terms of service, and whether we got away with it depended on whether the platforms felt inclined to enforce their rules. Plus, we heard rumours that Facebook — which had bought Instagram two years ago — could clamp down on the API any day now; we'd heard about companies being built on top of Facebook's API having their access turned off as a result of alleged data misuse, killing their entire business model. (Believe it or not, this was before Cambridge Analytica — the industry was rife with companies of that nature, and few ever batted an eye. If we had known about Cambridge Analytica's strategy at the time, rather than being horrified, we probably would have been jealous that we didn't come up with it first.)

The whole situation made me uncomfortable, and I suggested we find a new business model. Nick was less bothered — he didn't want to pivot just when we'd finally found some inklings of product-market fit. The platforms would only take action against us if they found out, Nick said, and why would they find out? And if they did find out, we would angle for an easy acquisition. Ask for forgiveness, not permission. Nick could be highly persuasive; we decided to return to building our product as if this existential threat did not exist.

By this point, Nick had found us a new niche. Our new arch-nemesis was a once-popular tech company whose briefly viral consumer product turned out to be a flash in the pan, with usage soon waning after the initial spike. But the company had raised enough money on the basis of its peak popularity that they couldn't easily pivot to something else; instead, they sold to a marketing technology company for a tidy sum. Under new ownership, this company monetised its former clout by selling user data for marketing purposes. It seemed like a shady outcome to me — it certainly wasn't the sort of company I wanted us to become — but their business model happened to be almost word for word what Nick had written in our pitch deck. Every time I looked over at Nick's screen, he was engrossed in their marketing materials or interviews with their CEO.

If we wanted to compete with this company, we would need help. Nick had been trying to get us advisors who would open doors and help us get sales in exchange for a small number of shares in our company. It was a fairly common strategy for startups in our space, but we were pivoting so frequently that it didn't quite work for us — if Nick found someone on Monday, by Friday we'd usually pivoted away from that sector so that it no longer made sense, and the advisor share paperwork we'd asked the lawyers to draw up would be left unfiled. Liam quipped that our cap table would end up like *Game of Thrones*: lots of characters, most of them killed off.

At least our pitch was improving, even if our optimistic projections about the multi-billion-dollar market we were capturing were quite a way off from reality. Sometimes the accelerator invited external guests to attend the pitch sessions and give feedback, and most of the feedback we got highlighted that our space was crowded and that we needed to differentiate. Yeah, thanks, we fucking know; we already read every word of glowing press written about our better-funded competitors.

I was itching to get real feedback, the sort that I didn't want to hear but needed to: that our idea was meaningless, that this whole startup dream was folly, that I did not need to be throwing away my twenties in pursuit of this thoughtless venture. But of course a startup accelerator was not going to host anyone who might undermine its raison d'être, and so there I was, stuck in a startup I didn't care about. Sometimes I thought I didn't need to care about it — sometimes I thought other founders didn't really care about their startups, either. It wasn't something I felt like I could talk to anyone about; everyone was always putting on a cheerful air, and it was hard to know whether the cheerfulness was genuine or not.

In a strange way, even though accelerators were meant to be supportive environments, I felt like we had to compete with the other founders. I knew the stats: for every ten startups, all but one would fail,[2] and I was determined that of the ten startups in our batch, the statistical outlier would be ours. I wasn't sure how we would get there, but I was positive that we could brute-force it by working as hard as possible, which I started to see as a badge of honour. The founders who never came in on weekends were clearly not dedicated; they would fail within the year.

Without ever explicitly discussing it, we started to ramp up our office presence. Tim wasn't always in on the weekends, and one Sunday I messaged him to ask where he was. He seemed annoyed by my unsubtle attempts to police his attendance, responding that he preferred to take the day off on Sundays and only work a half-day on Saturdays. My alarmed reaction was to gather an emergency meeting under the banner of team dynamics where I gave Tim plenty of opportunities to apologise for not working as hard as the rest of us. He didn't bite. Liam, Nick and I convened afterward, concerned about the growing disparity between our work hours, unsure what to do about it.

In place of confronting the deeper rift that opened up between us, we had proxy wars, and soon there was a muted tension underpinning even the most mundane interactions.

Tim thought we should look for a nicer place in a better neighbourhood, even if it meant a longer commute, but I was almost proud of the shoddiness of our current place and argued that a shorter commute was more important given that *some* of us worked seven days a week. A few of our fellow founders were even scrappier than we were, living out of suitcases and staying in discount hotels each night; the four of us living in a two-bedroom apartment was positively luxurious in comparison, even if Nick was consigned to sleeping on a fetid couch that gave him a rash.

In my mind, we had nothing but our youth. Our startup's mission was a battle, and youth was our most lethal weapon — hell, Mark Zuckerberg once said that young people were just smarter, and look how far he got. It would be silly to *not* exhaust our intellectual firepower in the service of crushing our enemies. Our youth was what gave us the ability to iterate faster, work harder, and innovate more than our competitors; our youth was what made us better than them. Better than the other founders in our accelerator, too. (When you're in the trenches, everyone on the outside looks like the enemy.)

It wasn't like anyone had flat-out *commanded* us to work this hard. Sure, the startup community normalised eighty-hour work weeks, yet there were plenty of rebuttals making the case for a healthy work-life balance. But I had come to believe strongly in the transformative power of hard work, not just as an obstacle to get through but as a key part of the process. That was my main takeaway from reading Elon Musk's biography — as long as you worked hard, you would figure it out. After all, Musk's first startup had a business model that sounded kind of boring, but it got acquired in an exit that made him an instant multi-millionaire, and now he was single-handedly saving the planet with Tesla and SpaceX. We could get there too, if only Tim would start coming in on Sundays.

Even as our team slowly started to come undone at the seams, I was still engrossed in the technical problems. One afternoon, I came up with an idea for linking Twitter and Instagram profiles together and making that part of our data offering. It was ethically a little iffy, as it wasn't data that anyone had exactly *consented* to give us, but still, the data was already out there, just itching to be used. (In those pre-Cambridge Analytica days, user privacy wasn't really a big concern in our sector.) After a week of frenzied prototyping, I hit the jackpot: we could link people by looking at their friends. Once I got that up and running, the number of links we could identify exploded, and all the charts started to go up and to the right.

While I was building this new feature, the others were running sales for our existing products. Our strategy had shifted into running reports on large consumer brands. We knew that we would lose most of these contracts to the smooth-talking sales teams of our better-funded competitors; still, I frequently stayed at the office past midnight doing free speculative work for multinational corporations in the hopes that one of them would take a chance with us. One company, a pioneer in the burgeoning subscription box trend, wanted an analysis of the different clusters in their social media audience. This process required a lot of manual intervention, but I was sure that if we kept tweaking it, eventually we would get to a point where minimal human supervision was needed. At least, that was the theory; truth be told, I was getting fed up with all these brand reports.

It soon got worse. One customer told us they wanted our data for a contract working, indirectly, with the US Republican Party. A brief team discussion concluded that even if some of us were personally opposed to the Republican Party, we should try to get this contract anyway — we needed the money. I manually confirmed every single piece of data we'd inferred for the users they sent over, just in case it helped the chances of getting the contract. I didn't really like the idea of working with the Republican Party, but if they were willing to pay, what else could we do? In war, you had to make alliances. In any case, this was just one piece of our powerful, complex, innovative system; there were tons of other use cases, which we just hadn't found yet. In the meantime, we would do free work even for the Republican Party if it meant more customer logos for our pitch deck. It wasn't our job to police our customers, after all — that would be time-consuming and internally divisive, not to mention financially ruinous.

It was easier if I didn't think about it too much. Don't worry about the end goal — just validate users, tweak parameters, optimise runtime. If I gritted my teeth and saw this through, eventually I would come out the other side where it would all be worth it.

One prospective customer, whose product was a platform for selling novelty T-shirts, wanted us to segment their followers. The whole process was very unscientific; after a lot of trial and error, we identified a significant minority interested in what we euphemistically called "country culture", which the company was actually surprised to learn about. We were elated: finally, a useful insight! We put this case study in our pitch deck and told the story verbatim to anyone who asked. It remained the only real success story of our reports feature.

One day, Liam declared that he was sick of running tests for brands. He thought we needed a better business model, like a self-serve API aimed at other startups. I agreed, and tried not to think about the possibility that these other startups might use our API to run tests for brands themselves.

I was sick, too — physically. One chilly September morning I woke up at 6am, my head feeling like it had been run through a wood chipper, unable to sleep through the bleating foghorn of trucks mingling with the high-pitched fury of cars on the freeway. There was nowhere better to go, so I dragged my exhausted body over to the office, where I found Nick dozing off on a beanbag.

After three months of this gruelling pace of work, we were running low on steam. Tim still wasn't coming in on Sundays, and we didn't know how to bring up our frustrations with him, perhaps out of a subconscious fear that we were the bad guys here. Team meetings were getting less productive — we were less patient, less willing to have the difficult but important arguments. I would take minutes anyway and delete the more contentious personal commentary that had accidentally slipped in before sending them around.

Our September partner meeting went surprisingly well given the state of our team dynamics. The partners wanted us to stop worrying about the specifics of what we were going to build and instead focus on our mission; they believed we would figure out the details along the way. That would have been reassuring, except we'd already been thinking that for over a year now, and it felt like we were running out of time before our founding team fell apart. Coincidentally, right after the meeting, Nick had a call with the former CTO of a company that did almost exactly what we did, and which failed precisely because their founding team fell apart. Ominous.

By the end of September, the situation with Tim was starting to become a real problem. Liam, Nick and I had decided that he wasn't pushing himself enough, and that — maybe even worse — he didn't care enough about startup culture. In short, he didn't fit our definition of what a founder should be. Bad culture fit, bad team dynamics, all around bad.

One Saturday, Tim was still awake by the time we returned from the office, and I determinedly convened a team meeting in the faintly lit living room. As Tim begrudgingly took a seat in the solitary chair, I detailed our central grievance: we thought he wasn't working hard enough.

From outside trickled in the sounds of rushing traffic, the occasional truck backing up into the parking lot across the street. Tim looked at me inscrutably for a few seconds before responding that he didn't understand why we felt the need to work so hard in the first place. He added that he was already working seventy hours a week — almost the equivalent of two full-time jobs.

I stared at him blankly, unsure how to respond. The concept of a full-time job wasn't in my vocabulary, because to me, we weren't really workers — we were pioneers, in a whole other orbit to something as quotidian as a job. I exchanged a mystified glance with Nick and Liam on the plastic-wrapped couch, who took turns explaining: Silicon Valley history practically had an altar to the eighty-hour week, and we thought that our hours were eminently reasonable for our stage.

Tim countered that most of the founders in our program didn't work as much as we did. That was true, but whereas Tim saw the other founders as the baseline, we saw them as mere occasions for us to reach personal bests. We thought we were just better at being founders, and we felt that our demands for Tim to ascend to our level were eminently reasonable. I tried another tack: yes, it might be a lot of work, but wouldn't you want to look back and be proud of how hard you worked? (Even now, far away from that life, it is profoundly discomforting to remember how much I believed my own bullshit.)

Another truck backed up, halting traffic on the street for half a minute; a predictable symphony of bleating horns followed. Tim blinked. He said he didn't think this was sustainable, us working this hard. He thought we were over-exerting ourselves for no real reason.

How to explain. Overworking, work for work's sake, crafting an entire identity around work: this was not a temporary anomaly, but rather my entire worldview. Working hard was an end in itself, and even those who chose not to do it had to realise that they were in the wrong, that their inability to work hard was a personal failing. The ability to achieve things through hard work was fundamental to my self-image, and this belief was buried so deeply in my psyche that I couldn't even identify it, much less relinquish it.

I looked at Tim, then looked away. (What do you say to someone when you're convinced that their fundamental axioms are wrong, and yours are right?) I restated my perspective: we were happy to throw all our effort into this on the off-chance that it worked out; I entreated Tim to do the same. He shrugged noncommittally, and as the conversation trailed off, we retreated to our rooms. I went to bed convinced that Tim would eventually come around and see things from our point of view.

I was twenty-three years old at this point, with a bottomless well of energy and drive, and the concept of ageing was a dim possibility on the distant horizon that rarely flitted into my peripheral vision. I could not conceive of a world where I no longer had the desire to pull all-nighters doing grunt work for a company whose mission I did not believe in, hankering after success as defined through a bricolage of authoritative words by wealthy men. I felt in my bones that hard work would save us, because this particular form of work contained within it the possibility of transcending work-as-usual. I had gotten a glimpse of the tedium of regular work at Google, and the grey dullness of it had frightened me so much that I had turned to this startup as a form of tacit escape.

There were all sorts of reasons we believed we could not afford to consider working less: our competitors were better funded; the industry moved quickly; we had grand ambitions. It was much harder to admit the real reason, which was that we were seeking something bigger, something maybe unfindable — that we were overworking not because we were doing well, but because we were doing badly. We were treading water in the desperate hope that someone would throw us a lifesaver.

The truth was, I was afraid to work less because it would make my life feel meaningless. I had wholeheartedly absorbed the belief that work should be the centre of my life, and yet the work I had found myself doing was nowhere near fulfilling. If this had been my real job, a 9-5 with OKRs and performance reviews, I would quit in an instant. The work was a slog, the use cases were trivial, and the customers annoyed me. But I had come too far to quit, so instead I accelerated forward.

Demo Day was coming up, and we needed to whittle our five-minute pitch down to perfection. I divided my to-do list into things which had to be done before Demo Day and things which could be done after. We weren't in great shape, and I thought Nick might be secretly panicking. Sometimes I snuck a peek at his laptop screen, only to see a blank slide.

Demo Day was meant to be the culmination of our four-month accelerator program, so it was a big deal, and we had all spent the summer tweaking our pitch decks. Your deck was supposed to put forward the best possible version of your startup — kind of like an online dating profile — but it seemed to me that the more it got refined, the further it drifted from reality. Your founding team had to sound like the smartest group of people to ever walk this Earth, so the team page should be stacked with references to institutions like MIT and Stanford and NASA. Most of all, your pitch had to be ambitious. Nobody wanted to invest in a company with 3% margins that would top out at a total addressable market of $10m — you might as well be selling sandwiches. Go big or go home.[3] Your potential market should be worth at least a billion, and you should be well-positioned to grab a big chunk of it.[4] It was crucial to show a hockey stick-shaped graph of your growth so far — revenue could come later — even if the numbers were artificially inflated. Hell, you could just make the numbers up. You wouldn't be lying, merely presenting the facts in the most flattering light.

Our pitch decks were not meant to describe reality; they created reality. Their goal was to sell a vision: the company described in these slides *could*, with the right investment, disrupt this industry and capture a slice of this multi-billion-dollar market. They were marketing, not fact, and their goal was to bring about the reality they described. And in a larger sense, those decks were selling the startup dream to outsiders — a dream to justify the exorbitant valuations. Startups that raised tons of money must have been deserving of that money, because their founders were brilliant and their aims were ambitious. Of *course* it would be a great idea to shower money onto smart people in an environment of minimal accountability. What could go wrong?

Demo Day arrived. We woke up early to prepare our booth. Nick was last to present. He did a surprisingly good job, and for a moment I really believed in the company — or at least, the lofty vision described in the five-minute pitch. As the last investors trickled out, we sighed in relief and changed out of our company T-shirts. The partners had rented out a karaoke bar; after several hours of day drinking, Liam and Nick took the stage to offer a horribly inebriated rendition of Eminem's "Lose Yourself", made worse when Liam fell asleep midway through. One of the partners caught the whole thing on video, and the next day he sent the link around the office.

With Demo Day over, it was time to get to everything we had put aside. Priority number one: move to a better apartment. The skylight mould in our current place was starting to block out the sun, and even I conceded that the place was intolerable. We found an Airbnb much farther away but quieter, more spacious, and with no discernible mould.

The office wasn't heated on the weekends, and after a long day of work the autumn chill lingered in the bones. One Sunday, finding ourselves alone in the office, we purloined a toaster oven from the communal kitchen and holed up in a conference room. In our toasty little grotto, we hammered out a six-month roadmap and brainstormed ways to build a moat around our data: we could trade it for e-commerce data to build a dataset that competitors couldn't reproduce! We could integrate with other networks! We could create a viral social app and monetise it later!

(Always trying to build a moat. Always treating data as something to trade even if we never got anyone's consent to use it in the first place. Ask for forgiveness, not permission.)

Mid-October, Twitter held its annual developer conference, and Jack Dorsey, who had recently returned to his former role of CEO, had some announcements to make. The good news was, they wanted a better relationship with developers who used their API. The bad news was, they would be releasing a product which competed directly with one of ours. Not for the first time, it occurred to me how silly it was to work on something that was essentially identical to what Twitter was building in-house.

But I was still working on it, and we were still in New York, even though we had originally planned to leave right after Demo Day. While we were here, we might as well make the most of it. We'd been burning through our Amazon Web Services (AWS) server credits, and an opportunity came up to get free credits in exchange for doing a bit of marketing for them. At the AWS developer pop-up loft in SoHo, I gave a five-minute presentation explaining our tech stack while profusely praising Amazon for their support; afterward, we got another $10,000 USD in server credits, as well as a swag bag containing an AWS hoodie in men's large and a Kindle Fire, among other things. Shilling for big corporations was nice, though Liam claimed my hoodie on the grounds that it fit him better.

In November, a massive advertising technology conference was scheduled to take place in the city. We showed up with lanyards and shuffled between monotonous talks on building customer profiles or IBM's UBX ecosystem, robotically collecting business cards from anyone we might want to integrate with or sell to. One of our biggest competitors had a booth here, and Liam and I decided to do some sleuthing. We casually walked up to the booth and asked the representative some probing questions about their product; she looked at us unblinkingly and asked to scan our badges. We gaped at each other in panicked silence for a few seconds before Liam walked away, pretending to take a phone call. I trailed after him, mouthing an apology. Some sleuths.

Meanwhile, the fundraising tour continued. Our latest pitch deck declared our unparalleled genius in supplying customer data to millennial-focused brands. Several potential investors told us no: the space was too crowded, they didn't have enough expertise in our area, we were too early. I told myself that we didn't want them anyway, and in any case, we had bigger problems to deal with. My latest diary entries were filled with agonised lamentations about the state of our team dynamics.

At least Liam, Nick and I were solid, drawn closer through our shared concerns about Tim. He usually left the office before we did, and the rest of us periodically took the opportunity to grab dinner on our own at whatever nearby restaurant was still open by the time we left. One evening, as we drowned our sorrows in endless shrimp at the Times Square Red Lobster, I wondered out loud if we might be better off without him. Liam and Nick exchanged triumphant glances and confessed they'd been waiting for me to admit this for a long time.

Mainly we felt that we did not understand him. We didn't share his need to have a life outside the company, and we were baffled by his numerous personal obligations to friends and family which seemed to us like a distraction; we were convinced that we would be a rocket ship if it weren't for his bizarre attachment to the idea of work-life balance. But as much as we wanted him to leave, none of us had the courage to actually initiate that conversation, so we continued working with him lackadaisically while secretly praying for a miracle. One of our partners had been dropping hints about acquiring us, and the more our team disintegrated, the more tempting it sounded.

Late November. The city was cold and damp and dirty. A new batch of founders was about to arrive, and the accelerator partners strongly hinted that we should get out of the office to make room for them. Under a heavy cloud of retreat, we packed our things.

SIX: CRASH

A startup is the largest endeavor over which you can have definite mastery. You can have agency not just over your own life, but over a small and important part of the world.
— Peter Thiel, co-founder of PayPal and partner at San Francisco venture capital Founders Fund[1]

Late November in Montreal was the cusp of winter: rainy days, temperatures dipping to below freezing at night, intermittent bursts of snow that quickly faded into the dull grey of concrete. We had thought that we would return with investors, customers, and a clear path forward. Instead, the summer produced no new investors and only minor progress with customers; most of our work lay abandoned in the form of half-finished proofs of concept from various pivots. Our broken team spirit was the elephant in the room, a festering open wound we didn't know how to treat.

In the brief interludes when everything had been going well, we had felt like a real team; the things that annoyed us about each other vanished into the background when we were winning. But whenever the situation darkened to unbearableness, the camaraderie would dissolve. We held endless team meetings where we circled around the fact of our unhealthy team dynamics, not realising that these aimless meetings were part of the very problem they were ostensibly trying to address. We were disgruntled soldiers cooped up in a trench far away from the front lines, and in the absence of a visible external enemy, we were quick to turn on each other.

At our next team meeting, Nick tried to unify us around the common enemy of the outside world. He envisioned for us a heroic marathon of outbound sales, and in anticipation of long enterprise sales cycles, he created a colour-coded spreadsheet estimating the number of cold emails we would need to send each day in order to reach fifteen closed sales by next June. This was his version of a motivational speech: he thought we had finally figured out our product, and now all we needed to do was fill up our sales funnel. If we put enough time and energy into this problem, we could overcome it. Tim looked a little unconvinced, but I had faith that if we threw enough at the meta-problem of convincing him, we could accomplish that, too.

Fall sloped into winter. After the speed and urgency of the summer, everything suddenly shifted into hibernation. Sending a hundred cold emails each day turned out to be a lot harder than it sounded, and as we started to exhaust the pool of potential customers, our energy began to dissipate. Liam and I began playing RuneScape again: work had become woefully unsatisfying, and it was nice to have an immediately gratifying outlet for our energy. The game's levelling-up mechanics were much more straightforward than our meagre sales funnel, and I needed to briefly suspend myself in the illusion that our daily grind would add up to a transcendent whole.

In December, we got a term sheet from a New York-based venture capital firm. It was our first real offer of funding, and it came with a substantial amount of money at a mildly disappointing valuation. We dithered about it at length; we weren't excited by the firm's portfolio, and the terms seemed onerous. Apparently this was what it meant to be your own boss: either you barely had an impact because you had no resources, or you had to raise money in a way that gave control of the reins to someone else. We declined the offer.

In the midst of this fundraising paralysis, getting acquired started to look like a decent alternative. One of our partners offered to fly us out to their US headquarters, ostensibly to hammer out partnership details, but really, we suspected, to sweet-talk us into an acquisition. They were headquartered in a mid-size city in a state I had never visited before, but I was ready to give myself away to them if that would resolve our lack of progress. I browsed apartments listings near their office, trying to imagine myself working for this company whose product I didn't care about. At least it couldn't be worse than working for my own company whose product I didn't care about.

In the weeks leading up to the trip across the border, our attitudes fluctuated wildly. When things were going well, we celebrated milestones by reminiscing about our journey over all-you-can-eat sushi; Nick suggested that we should hold out for eight million, while Tim said that he wouldn't want to sell even if they offered us twenty. Other days, when we could barely stand to look at each other, I pinned my hopes on the escape symbolised by selling.

By the time we arrived at their headquarters in January 2016, we still hadn't settled on whether we wanted this acquisition or not. We spent hours upon hours in a conference room going through financial projections and product roadmaps with their executives under a strict non-disclosure agreement. In the middle of a long technical debate, their CEO stepped out of the room, motioning for Nick to join him; they stepped back in a few minutes later, their faces inscrutable, the rest of us pretending not to notice.

On the flight back home, Nick filled us in. "He wanted me to name our number first, so I said eight. He said, wow, that's high, so I said, I know you're not going to give me what

I *ask* for. He laughed, so maybe we're good." I thought that number was absurdly high, but maybe that was just how negotiation worked. We threw out guesses for what their counter-offer might be: six million? Four million? What was our threshold? Those numbers were all so arbitrary anyway, completely divorced from any material reality; in this industry, acquisition amounts were usually based on gut feeling more than actual metrics. Wisps of cloud trailed lazily underneath our airplane.

I thought about what the acquisition would mean for me personally. I didn't have any debt, and my small paycheck was enough to cover the basics; I wasn't sure what I'd need the money for. But the value of the money was less in its purchasing power and more in what it signified: success, validation, negotiating with the executive suite rather than being just another salaried employee... and after over a year of being a founder, I'd gotten addicted to this feeling. The money would be nice if I ever decided to do another startup, but its main function would be symbolic, not fiscal.

Unbidden, a statistic came to mind — something I had read recently about many Americans not being able to come up with $400 in an emergency.[2] The realities of economic hardship for the average person felt distant from where I was, hurtling through the rarefied tropospheric air, and at first glance seemed unrelated to the million-dollar acquisition we thought was coming our way. And yet, at the end of the day, it was the same system of money. It was a system that functioned both as a gatekeeper for ensuring survival, and as a means for people like me to demonstrate value. It briefly occurred to me that it might be unreasonable to expect it to do both.

A week later, our partner sent their main negotiator to Montreal to deliver their counter-offer in person. It was the barest whisper of an offer; we were told that the most they could do was one — the word "million" was never actually said, only intimated — split 50/50 between cash and stock, over three years, on top of market-rate salaries. They would help us relocate to the US, and we would keep working on the same product, only as a small division of their company.

We tried to hide our disappointment. Was this a standard negotiating tactic on their part, or was this their way of telling us to fuck off? The fact that they communicated this in person told us that they were genuinely invested in making it work. Perhaps they really couldn't afford more. But we weren't thinking in terms of what they could afford — we were looking outward, at other startups, some of which had sold for way more on the basis of far less. I felt blindsided by what I saw as a massive lowball; my reaction bordered on anger. Sure, these other valuations might have been symptomatic of a tech bubble rather than indications of real value, but on some level we believed that the valuations were justified — we wouldn't be doing all this if we didn't.

The funny thing was, I knew I wasn't currently motivated to work by the money, and yet the idea of being offered a market-rate salary and a slightly boosted stock grant made me feel cheated. I knew that was silly; it was hard to reason about this in a way that didn't seem greedy. In search of a comparison, I calculated how much I would have made if I had gone to Google, treating that as a minimum for how much I would have to get in an acquisition. It was a narcissistic distraction, of course, since I couldn't go back in time to make that choice, but it was the only barometer I had.

In the end, we rejected the offer on the grounds of valuing our independence. Really, though, it was because we thought we were worth more than their assessment of us, and given that we didn't really believe in our mission, the pale prospect of future financial validation was all we had.

Once we turned down the offer, it was back to the grind — no more fantasising about moving to the US. We were pivoting in circles again, our attention jumping from idea to idea. Tim was trying to get us political contracts related to the upcoming US presidential election; Liam and Nick were working on a contract with the email marketing department of a large consumer brand; I was building integrations with startups that helped Instagram influencers to monetise their audiences. The days passed by in a grey haze of data dumps and rote maintenance work, and I stayed up late less because I was excited about the work and more out of a half-hearted attempt to postpone tomorrow. Just another thousand rows of data and I would go to bed.

If this were a job I did only for the paycheck and the rest of my life were elsewhere, then even if it felt dull, at least it would be understandable. But this whole endeavour was an attempt to *avoid* the drudgery of working for someone else — an attempt to define work on my own terms, to make it exactly what I wanted it to be. Whatever mark I might leave on this world was mediated by my work, and I was growing increasingly worried that this work was not especially meaningful.

I had believed that I could escape. I set out for this startup in the hopes that it would be a chance to be my own boss: to work on my schedule and on my own terms, rather than being shackled to a 9-5 with OKRs and a manager. And yet whatever freedom I thought I would have turned out to be illusory — my startup was still governed by the same market forces, and it was starting to look like there was no real freedom to be found within this system. The lights were blinking back on, and for the first time I noticed that the doors around us all had padlocks on them.

The accomplishments I used to crave seemed less

and less promising. I no longer wanted revenue, growth, glowing write-ups in the tech press; I just wanted to feel like it all meant something. I wanted to return to the beginning, to the metaphorical garage, to that first summer when we didn't have to worry about anything other than building something new and exciting. I wanted to return to the thrill of creating something from scratch, back when the pressure of actually selling our product was a distant twinkle on the horizon.

I felt like I was choking. I had thought that this was the path I was supposed to follow — the fabled entrepreneurial journey — and yet everything about it felt hollow. But even as I was losing hope in the possibility of reaching the promised land, I still wasn't ready to give up. It would be too horrifying to admit that I'd made a mistake when I didn't know what the right answer would be.

By March, the walls were closing in on us. The platforms we got our data from announced that they would be shutting off the endpoints we relied on, essentially asphyxiating our business model. From our perspective, after having spent nearly two years building our technology, this felt like rank unfairness, even though the average user of these platforms would probably be offended that companies like ours existed at all. In a way, it was a relief: we no longer had to worry about the worst happening, because it had already happened. Our commitment to loading up our sales funnel dwindled, and our colour-coded spreadsheet slipped further and further out of date.

Once we had a death sentence on our business model, the possibility of giving up began to lurk overhead like a gloomy cloud. As our planned roadmap fell by the wayside, Liam, Nick and I started reflecting on what we had learned. We boiled it down to the three most important lessons: choose your co-founders carefully; avoid platform risk;

understand the market before building the product. ("Build a product you actually care about" wasn't yet on our radar.)

The funny thing was, whenever we had to explain the Tim situation to other founders, they got it instantly: we needed a hustler — someone eager to work as hard as we were — and he wasn't one. Outside the context of startup culture, it would probably sound like unhealthy workplace expectations; in my mind, though, it was cut and dried. In my diary, I wrote pages and pages of justifications addressed to an imaginary jury. In my recounting, I was always on the good side, the side of the hungry and soon-to-be-successful founder, able to disrupt whole industries through grit and passion; on the other side was Tim, with his baffling lack of commitment to the cause.

I proposed countless ideas for "correcting" Tim's attitude. One team exercise where we were supposed to brainstorm our corporate values was a total setup, because Liam, Nick and I had agreed on the words ahead of time; we just wanted to know what Tim would say. I threw out "lean", expecting it to be an easy one — we'd all read Eric Ries' seminal book *Lean Startup*, and we even had two copies in our office, the extra copy being perhaps ironically superfluous yet symbolic — but Tim bristled. He retorted that we were glamorising efficiency and leanness to the point of absurdity. I was taken aback, and I suggested (good Lord) that he read *Lean Startup* again.

After that, the atmosphere — already on a knife-edge — soured beyond repair. Days passed by in near silence, with in-person exchanges limited to perfunctory greetings; work conversations took place primarily over Slack. I sent Tim a wall of text where I explained that we would all feel better if he made an effort to familiarise himself with startup culture; he didn't respond. The rest of us created a separate Slack channel dedicated to surveying Tim's work ethic and sharing articles about how to deal with co-founder incompatibilities.

By the end of March, it was brutally clear that our business model had no future. Liam, Nick and I were leaning towards selling it off to the highest bidder. That was the easy part; the hard part was telling Tim that we wanted to take another shot at the startup goal, but with just the three of us. I went for a walk with Tim where I explained this as delicately as possible, though still framed in the hypothetical.

Mid-April, we decided. We couldn't work with him anymore — the shared goodwill so crucial to any founding team was gone. We finally mustered enough courage to initiate the final conversation, which Tim seemed to have seen coming; he responded more graciously than we probably deserved. And that was the end of that.

SEVEN: PIVOT

The only way that I can see to deploy this much financial resource is by converting my Amazon winnings into space travel. That is basically it.
— Amazon CEO Jeff Bezos in an interview with *Business Insider* in 2018, when his net worth was estimated at $131 billion[1]

Springtime in Montreal: soft dewy mornings, budding New World flowers, occasional scatters of warm rain. After a lethargic winter, we were slowly blinking awake again. Shorn of the one team member we thought was holding us back, we suddenly had a new burst of energy. Colours were flooding in; everything felt brighter, crisper, more melodic. The air was buzzing with possibility.

Now that we had come to terms with how much we hated our business model, we fixated on pivoting to something completely different. Founders did this all the time — Twitter, Slack, YouTube, and Yelp were all the result of successfully pivoting away from failing initial business models. This time, we would get it right; this time, we could do anything. We were older, wiser, better-connected. The world felt open to us in a way that it never did before.

Our work responsibilities were minimal given that we had decided to halt product development, so we had plenty of time to explore the blue-sky ideas that we wanted to be working on instead. Automation was high up on that list. One afternoon, Liam reverently showed us a video about Amazon's robotics division — proof of Amazon's efforts to automate away dangerous jobs, which we saw as a sign of the benevolence of Amazon CEO Jeff Bezos who clearly cared so much about making life better for his workers. Our perspectives were immature and half-formed, with no theoretical basis, but we expected we could figure everything out by reasoning from first principles, as if we were the first people to have ever thought about the concept. We devoured blog posts from Silicon Valley thought leaders on potential policy responses to widespread automation; I was especially intrigued by the idea of Universal Basic Income, an idea which I assumed had been invented in Silicon Valley.

I wasn't entirely sure what we could do, but I took solace in the fact that our existing product was a narrow application of the general idea of automation, in the sense of reducing human intervention using machine intelligence. I was sure we could adapt our technology for other use cases as well. We were smart, we had studied computer science, we had read *Zero to One* and *The Hard Thing About Hard Things* and *Freakonomics*; we could boil any problem down to a neat object-relational mapping abstraction.

Our idea of market research was Googling for various keywords related to automation. The startups we found all seemed to have narrow focuses, like scraping websites. My ambitions were much grander: I wanted to build a platform for automating human labour in a wide variety of verticals. As before, we would start with a particular beachhead market, then expand from there to be the technical layer underlying all kinds of labour-saving automation tech. (The idea might have been different this time, but just as with our previous business model, the goal was domination.)

This idea really fascinated me, so I turned it over in my mind, paying special attention to the ethical considerations. After the last two years of toiling away at a product with questionable social value, I needed my next startup to be the opposite — an unabashed social good I could feel virtuous about working on. In my diary I jotted down a tentative mission statement: "transition humanity to a state where human labour is valued and is harnessed to its full potential". That meant no more wasting human time on repetitive, easily-automatable tasks; instead, machines could be used to liberate people from necessity. Even beyond saving money and reducing error rates, I was confident that the true benefit of automation was giving humans back their time. I wasn't entirely sure how we could accomplish that, but I had faith that we would figure it out.

In the meantime, the next Demo Day of our erstwhile accelerator happened to coincide with my 24th birthday. We took a team trip to New York to recharge and catch up with the other founders from our batch. It was a little bittersweet to be at Demo Day again, seeing the latest batch of founders going through the meat grinder, knowing that most of them would get crushed.

A fellow founder, another alumni of the accelerator, invited us to stop by his Midtown office filled with floor-to-ceiling windows and tastefully expensive atelier furniture. For an hour, in a conference room surrounded by breath-taking views of the Manhattan skyline, we caught him up on everything that had happened: the co-founder breakup, the death of our business model, our plans to pivot into an automation company under the name "Turing Labs" (later abandoned on account of its association with Martin Shkreli's price-gouging pharmaceutical company[2]). He was alternatingly sympathetic and encouraging, and as we left, I felt a little lighter, a little more optimistic. After all, *this* startup had once been merely an idea, and now they were a wildly promising marketing technology company with panoramic Manhattan views and millions of dollars in funding. We could make it, too.

Last time, we had wanted to do a startup, any startup,

and so we had grabbed for the low-hanging fruit, only to discover it was already rotten. This time, though, we were going to do it right. This time we were going to build something we actually believed in.

After two years of working on our failed ad-tech play, we were finally, finally ready to give up. The actual mechanics of giving up were still unclear — we threw around terms like retiring, sunsetting, and pivoting almost interchangeably. Nick re-opened acquisition talks with companies that had made overtures in the past, beginning with offers to sell off pieces of our database. It was a little painful to have to say goodbye to a project I had devoted so much time to, but rationally I knew it was time to move on.

In May, we moved our office to a co-working space for startups where we prepared for inspiration to strike. Each day, once the urgent work had been cleared out of the way, it was time for blue-sky thinking. We devoured all the entrepreneurial motivation we could find: blog posts from founders who had failed but were trying again, pivoting how-tos with advice like "discuss things you're excited about". It was a little like group therapy for burnt-out startup founders, with the topics of discussions provided by *TechCrunch* headlines or VC Twitter: chatbot startups, self-driving cars, which latest tech luminary was accused of sexual harassment. Nick brought up health-tech, which Liam and I vetoed because we had no expertise in that field; I suggested something to do with food, despite having no expertise in that either. Liam babbled about fin-tech companies with the potential to democratise finance. As long as we kept talking, eventually an idea would take root.

Some days, I worried that we were fooling ourselves. There was no guarantee that we would find the right idea immediately, and the thought of more pivots on the horizon was unbearable. Other days, I could already see myself giving a keynote at some startup conference several years down the line about the heart-warming story behind our miraculous pivot.

Despite our former team-rending insistence on the necessity of hard work, our rhythm had gotten a little healthier: normal hours during the week, time off on the weekends. I started hanging out at a nearby park on Saturdays, reading Neal Stephenson novels or whatever links I came across on social media. One weekend I stumbled upon an *n+1* essay titled "Uncanny Valley" by Anna Wiener, shared by some tech people I followed on Twitter.[3] I had never heard of *n+1* before, which I assumed was a tech publication from the name, but soon learned was a literary magazine. The essay stitched together a series of vignettes about working in tech, and I was immediately engrossed, holding on to each line as it scrolled by; when I got to the end, I scrolled to the top and read it again. The words cut like a knife through my gradually waning hopes, and I wanted to sink into an ocean of this writing.

Even though I'd gotten more critical of the industry over the last few years, I still ultimately believed in it. I'd staked my identity on it, after all; I felt personally invested in its success. But reading and re-reading this essay, I started to see the industry in a less flattering light: my own startup journey began to look absurd, and our last-ditch attempt to pivot appeared delusional, a desperate attempt to remain in denial about our impending failure. But this essay's author had been in a similar boat, and even though her startup journey seemed to have a bittersweet ending, she had found a way to move on. A hopeful seed began to sprout in the edges of my mind: maybe I, too, could make it through this gloomy tunnel of endless pivoting. And maybe I would come out the other side a different person, able to look back on the dejection of our current moment with grace.

Most of all, I wanted to get to a place where all this ceased to hurt so much. Every failure was a whip lashing away at my increasingly fragile self-esteem, and even the successes seemed tiny compared to how far we needed to go. I wasn't sure that I could motivate myself much longer, but I hated myself for feeling this way, because it seemed like weakness; I despised weakness, most of all in myself. I wanted to be stronger than this. I wanted to pass this test of faith and emerge anew, shedding my self-hatred like an old skin.

May morphed into June, the warmer sun chasing away the dusk, us still chasing after the shadow of our elusive pivot. Our latest blue-sky idea involved automating away employee handbooks by storing worker directives as rules that could be interpreted by a machine, and we thought call centres might be a good place to start. (In other words, we were trying to codify the labour process in a large sector of the economy, in which none of us had worked, in order to automate away jobs and take some of the cost savings ourselves.) This wasn't exactly the socially-driven mission I had envisioned back in May, but the technical problem captivated me; I could worry about the ethics later.

Yet it was unclear where to go from the initial ideation step. We didn't have contacts in this space, and I'd never talked to a call centre worker outside business hours — I had no clue what their work was like. The novelty soon faded, and I realised I wasn't ready to devote the next several years of my life to working on this problem. The idea was quietly dropped, never to be spoken of again.

That was OK, though, because we had no shortage of ideas. The summer flew by in a whirl of blue-sky thinking. We debated, argued, occasionally even agreed; it was the most fun I'd had since my college hackathon days. Sometimes it was unclear if we were discussing things we wanted to personally work on, versus things we merely wished would exist. The word "postcapitalism" was bandied about a lot, inspired by the musings of an NYC-based venture capitalist who had recently started blogging about it[4]; we weren't entirely sure what the word meant, but we wanted to do a startup roughly along those lines. In our hazy mental map, Tesla seemed about as good a mascot for postcapitalism as any company: environmentally friendly, open-sourced patents, a brilliant CEO.

Our discussions were starting to border on light critiques of capitalism. Climate change kept popping up in the news that summer, and when we talked about it, it was with a recognition that ecological problems often had an economic cause — that environmental externality costs were simply not priced in under the current system. And advertising was my personal pet peeve Our time spent in the ad-tech space had convinced me that most of the industry was bloat, and I wanted to live in a world where the prevalence of advertising was radically diminished. I didn't like the idea of an industry that essentially manipulated people into buying things they didn't need in order to pad sales figures.

For our next company, we were going to do something meaningful, not simply technology for its own sake. We were going to build something that we cared about, that people wanted, that we could fund without raising a ton of capital. If we could just settle on an idea like that, we were golden.

In the meantime, as we went through the list of every possible blue-sky pivot, our current business was placed on life support. We started downsizing our tech infrastructure: making harsh cuts to memory and space usage, switching to on-demand server provision despite the potential slowdowns to our product, even shortening logging output to save a few extra dollars a month. Our Amazon Web Services credits had run out, and every dollar counted; the longer we could keep our existing business alive, the more time we had to figure out our improbable pivot.

Robotics had piqued my interest. I loved the idea of automating away rote tasks, and my latest obsession was restaurant work — something that I had, again, never done, but which I assumed I could optimise. My vision was an industrial robot that automated food production, so we could do away with pesky labour costs in order to sell food on the cheap. I did some research into the restaurant industry, only to learn that labour costs were already quite low — in much of the US, the minimum wage for tipped roles was a mere $2.13 an hour; the biggest expenses tended to be the cost of raw materials and rent. Rent was out of our control, and raw food costs would have to be addressed somewhere in the agricultural industry, though Nick got carried away on an ag-tech tangent for a few days and rambled effusively about putting sensors on tractors.

Clearly my initial assumption had been wrong — restaurant food was not expensive primarily because workers were being overpaid for something a machine could easily do. That put a small wrench into my plans, but I still thought we could find a niche in this market. In my experience, prepared food tended to meet at most two of the healthy/fast/cheap trifecta; what if we could meet them all?

It so happened that a fast-food restaurant specialising in quinoa bowls had just opened down the street. The next time I went, I did what I thought of as market research. As I watched the employee assembling my requested ingredients into a bowl, I pictured a machine resembling a soda fountain in her place. A smooth, frictionless self-serve system for combining ingredients, with ordering and payment handled through a mobile app that required minimal input from workers. Humans were expensive, requiring minimum wage, sick days, lunch breaks; sometimes they misheard what the customer asked for. Humanity was messy. Much better to have a machine of gleaming steel managed through an unambiguous codebase.

If I ran that startup, the employee currently assembling my quinoa bowl would be fired. Service workers were unproductive labour, the sort that ought to be optimised away; I would hire only high-performing engineers and the people needed to support them. If I needed service workers to do maintenance work or food prep, I would simply contract that out for the cheapest wages the market could bear — the innovation behind the recent spate of gig economy startups. Independent contractors wouldn't qualify for minimum wage protections or benefits, after all.

What would happen to the employee? Well, she could find another job. Of course, other jobs were getting automated or made more unpleasant through the intensification of surveillance technology, but the plight of this employee was not my problem. My only responsibility was to my technology, my investors, the higher notion of progress; the human beings at the short end of the stick were inconveniences to be automated away, necessary sacrifices at the altar of efficiency.

But I was not running that startup, so this employee got to keep her job, and the cost of my quinoa bowl went partly towards allowing her to pay her bills as opposed to being funnelled entirely to some Silicon Valley startup making a return for its investors. The engines of progress were stalled, and instead I would have to settle for the subpar status quo where people who weren't engineers were able to survive under capitalism.

OK, so "food robots" was apparently fraught with all sorts of thorny ethical issues, who knew? But what about self-driving trucks? The media had been asking what would happen to America's truck drivers when the work was all automated, and we eagerly brainstormed policies that might prevent potential riots. Nick proposed some sort of data dividend, with truck drivers being paid for any driving they'd done that contributed to the algorithm. I vetoed that instantly — if there was anything I learned from our prior business model, it was that data attribution was nearly impossible. Not only would it be technically inefficient, but it would erode the profit motive for the company doing the automating. In any case, there was no neutral way to assess the share of income that should go to drivers; it would be entirely dependent on their negotiating power. I didn't know the right way to solve this problem — a basic income? collectively-owned technology? — but the typical Silicon Valley approach didn't seem apt here. It felt like it would be better to pay drivers what they needed, rather than concocting some byzantine and arbitrary scheme for assessing their data contributions.

One day, I came up with an idea that seemed like it would be the answer to all our problems. The thing we hated about our previous business model was the industry; the tech was actually pretty cool. What if we could keep the tech, but change the business model? No more hobnobbing at ad-tech conferences — we could be a generic data platform, open to anyone who needed training data for machine learning. If this worked, it would be the quintessential pivot, and I was so excited by the possibility that I registered a domain and whipped up a splash website, even before the others agreed. But after a bit of research, I realised that the platforms we would integrate with for generating training data were teeming with ethical issues — their entire business model relied on a global army of underpaid workers who weren't covered by traditional labour laws. My enthusiasm faded.

After months of these discussions, I was starting to lose faith in what I was doing. The lean startup toolkit I'd been equipped with seemed increasingly inadequate for the problems I wanted to work on. Nick would poke holes in all of my proposals, which I found highly annoying, but maybe he was right — the startup model might not be appropriate for any of my blue-sky ideas for social good.

Maybe I had it backwards. I'd assumed that the presence of unfulfilling work was the problem, and so the solution would involve machines taking over that work. But the very presence of unfulfilling work raised the question of why people were doing those jobs at all. Unless we tackled that part, then automating those jobs away wouldn't necessarily give those workers their time back; instead, it would probably just get them fired, consigning them to a series of even worse jobs. Perhaps the real problem wasn't the technical problem of automating away jobs, but rather that we didn't have an economy where it was humane to focus solely on automation in the first place.

Late August, we flew out to San Francisco again. This time, we had no meetings lined up; it was solely for vacation. We made a day trip to Sequoia National Park, and walking beneath centuries-old Sierra redwoods was an oddly comforting reminder of the frivolity of my startup woes. The drive back to San Francisco passed by in silence, and as night fell, an eerie mist settled in, shrouding the silhouetted trees on either side.

Questions surrounded us, clinging to us like fog. Were we still going to pivot? How much longer could we do this? What if this whole startup dream was a scam? We could put them at bay for a while, but they were starting to close in on us.

We had really thought that, once Tim had left, we would be a unified team, and nothing would stop us. But after several months of blue-sky debates with just the three of us, the sky was still as cloudy as ever. We couldn't even agree on an idea that we all wanted to do. The central fissure had Nick on one side, Liam and I on the other. Nick styled himself as the voice of economic reason, raising objections to any ideas Liam and I proposed on the grounds that they wouldn't succeed as viable businesses, and really, he was right. On the other hand, Nick kept proposing ideas to do with the US healthcare system, citing examples of other ad-tech founders who had gone on to make big waves in that space, whereas Liam and I were opposed to private healthcare on principle.

I had wanted a cause I could wholeheartedly believe in, something as far away from ad-tech as possible. But all the socially useful ideas I wanted to work on were simply not economically feasible. And all the potentially lucrative ideas, the low-hanging fruit of startups, seemed at best boring and unnecessary, and at worst morally abhorrent. But I didn't know what came after a failed startup other than pivoting our way out of it. I'd read my fair share of Silicon Valley failure stories, and most of them ended with the founders dusting themselves off and getting up again. Failure was only acceptable when it led to something better, and I had staked my whole identity on being the kind of person who could achieve anything. But what if I couldn't?

By September of 2016, our runway wasn't looking so good. Our earliest customers were cancelling or downgrading their contracts. One of our partners had been acquired by another company and would no longer be needing our services. We were entirely out of web hosting credits, and merely keeping our servers running was draining our bank account. And our planned pivot — that fragile bubble of an idea — was disintegrating before our eyes.

The partners of the accelerator recommended that we go for an acquisition ASAP. They were disappointed that we were giving up on this business model, but it wasn't a big deal for them — they had only invested a small sum in our company, and they would probably get it back in the acquisition anyway. Plus, they expected most of their companies to fail; they would rather we get acquired or even shut down than run a three-person lifestyle business. If it wasn't growing, what was the point? We arranged calls with various Corporate Development departments, shopping around for a suitable acqui-hire — that is, a joint job offer thinly disguised as a product acquisition. If our product had been successful, we would expect to command maybe a million dollars per head, but given that we were slowly dying off, we would be lucky to get a million in total.

By October, I had almost entirely checked out of the startup world. Instead, I spent most of my time in the office following coverage of the upcoming US presidential election. I couldn't tear myself away; I didn't think that Donald Trump could actually win, but the possibility horrified me nonetheless. Our conversations dwindled and soon the hours ticked by in silence.

Even if we did find a pivot we could agree on, it hit me that I could not do another two years like the previous two. I no longer had the energy of being twenty-two and excited about startups, and the thought of enduring even another month of that grind felt like climbing an impossibly vertiginous mountain. I was so, so tired, and everything around me seemed to be drenched in a hopeless fog. I couldn't even remember what we were doing this for in the first place.

Nick was still on the hopeful side, but he could tell that Liam and I were slowly burning out. One day, out of the blue, he asked, "Are we still doing this pivot?" Liam and I looked at each other and shrugged. The promised land didn't look so promising anymore.

EIGHT: GIVING UP

But back I cannot go, this waste of time, this admission of having been on the wrong track would be unbearable for me. [...] The time allotted to you is so short that if you lose one second you have already lost your whole life, for it is no longer, it is always just as long as the time you lose. [...] As long as you don't stop climbing, the stairs won't end, under your climbing feet they will go on growing upwards.
— Franz Kafka, "The Departure"[1]

Summer faded into autumn. The days got shorter and crisper. Shadows lengthened as leaves turned crimson and drifted toward the ground. The realisation that I had been chasing the wrong thing was starting to catch up on me.

The thing is, every little thing we did had seemed rational at the time. We had begun with technology that we thought would disrupt an ossified market; from there, we had iterated our way around the sector in pursuit of product-market fit, repeating those trochaic syllables as if we were saying the rosary. Always pivoting just in time to avoid the raw emptiness at the edges of this ambition. Two years of aimless pivoting later, we had decided we didn't like our business model and wanted to try something new. Another startup idea, another shot at the goal, another knock at the hallowed gates of Silicon Valley success.

Yet after a summer of futile brainstorming, our hopes of finding a miracle pivot were deflating. But I didn't know what else I could put my hopes in.

And then, right in the middle of my personal startup pity

party, Donald Trump was elected President of the United States. I took a couple of days off to do what every other liberal seemed to be doing: mourning, decrying racism and sexism, scrolling endlessly through social media.

For years, our expectation had been to move to the US eventually, simply because that was where all the money was. At first, the San Francisco Bay Area had been our target; after spending a summer in NYC, we had switched our sights to the East Coast. But we had been counting on Hillary Clinton becoming President, on the grounds that she would be more likely to pass immigration reform that would make it easier for people like us to move. Never could we have imagined a President who would turn the clock back the other way, making an already restrictive immigration regime even more nativist. My longtime dream of moving to the US slipped further out of reach.

And in the meantime, I was watching to see how the tech industry would respond to this new administration. Silicon Valley was often stereotyped as liberal — at least socially progressive if sometimes fiscally conservative — and I expected high-profile tech leaders to use every lever at their disposal to put up resistance. I was left severely disappointed. There were a few bold statements in condemnation of Trump, but most of Silicon Valley seemed perfectly willing to work with his administration, gathering at the White House for roundtables[2] and sending surrogates to staff the transition team.[3] Election over, now back to business as usual.

But the tech industry had minted some of the world's richest people, many of whom claimed to actually care about making the world a better place. Call me naive, but I had genuinely believed them — I had really thought that tech would save us.

What was the point of all that wealth and power if they refused to use it when it mattered?

Few Silicon Valley luminaries had been public Trump supporters in the run-up to the election. One notable exception was Peter Thiel, PayPal co-founder turned prolific Silicon Valley investor, who had donated $1.25m to Trump's campaign and had spoken in support of him at the Republican National Convention. Thiel was a part-time partner at Y Combinator, and even though fund president Sam Altman had been vocal in his opposition to Trump, he had refused calls to remove Thiel from the fund.[4] I couldn't really fault him: Thiel was deeply integrated into the Silicon Valley investment circuit, and Y Combinator couldn't afford to alienate him. On the other hand, there was a larger concern here: if those who actually had leverage over people like Thiel were too afraid to use it, what were the rest of us supposed to do? What were the feedback mechanisms if those with power were unwilling to exercise it on behalf of those who *didn't* have power?

For a while, I avidly watched the same liberal late-night talk shows as before, following the hosts as they switched networks or started new shows. Soon, the jokes began to feel stale — there were only so many ways you could insult the president's appearance or incivility before it seemed like that was all you had. The liberal perspective I had been accustomed to just wasn't doing it for me anymore, as I was losing my faith that the establishment would set things right again. Eventually I tuned out altogether.

Once Trump was elected, my startup aspirations hit rock bottom. I wasn't sure what I should be doing instead, but now that I was finally paying attention to what was happening in the world around me, it felt frivolous to go back to what I was doing before.

Some of my friends were in a similar state of post-Trump

malaise, suddenly uninspired about pursuing their previous career goals. One friend had recently switched careers to join a non-profit advocating for prison reform. I wondered if I should be doing something less self-aggrandising and more relevant to the current political landscape.

As the world grew darker and I began to seriously consider alternative life paths, my uneasiness with what I'd been doing for the past two years started to catch up to me. Liam and I had come to this realisation more or less simultaneously, and we spent countless hours discussing why we had done this for so long. Why had we thought that it was OK to sell personal data without consent so that brands could target people better? I looked back through my earliest diary entries, and it was almost physically painful to see what I chose to record on a day-by-day basis: the number of users we tracked, the attributes we could infer, the meticulous technical details of the bugs I fixed — all because I had really believed that it would matter. I had thought that I would look back one day from the height of success and write my own version of *The Hard Thing About Hard Things*.

The thing is, I had always suspected, deep down, that what we were doing was wrong. Our business model had never felt particularly respectable, nor did it feel like it contributed to society in a way that legitimated even the little revenue we got. Whenever I had to explain what my startup did, I usually added the caveat that the business model was a bit boring, but that the technical problems were fun; who would question that?

No one had ever told us that what we were doing might be wrong. The criticism we did get would mention the platform risk we were incurring, the crowdedness of the space, or the potential commoditisation of our technology. The people on whom we counted for advice or money or connections never raised an ethical objection to our flagrant disregard for other people's privacy. Why would they? We had seen a niche in the market, and we tried to fill it. The market was the ultimate arbiter. Questions of ethics were beside the point.

Whatever doubts might have lurked in the back of my mind always paled in comparison to the solidity of the larger ecosystem. After all, we were a tiny cog in a much larger machine. Everyone we dealt with knew exactly what we were doing and still dealt with us anyway. Respectable companies and individuals wanted to invest in us, acquire us, trot us out to promote their cloud computing services; how could we not have seen that as validation? The whole system had felt so *sturdy*. I couldn't have doubts about my startup without the whole thread unravelling, threatening the foundation of my entire belief system.

But now the thread was starting to unravel, and the larger tech industry wasn't doing much to settle my doubts. How many more unflattering stories did I need to hear before I lost faith? My belief in the tech sector was starting to look less like a neutral assessment of merit and more like a tale spun for self-preservation. I had wanted so badly to believe that I was special, and when I had found a refuge that granted me what I wanted, I clung on to it, convinced that it was deserved: people like me, who were good at manipulating computers, were finally getting justice from a world that had previously failed to validate them. The stagnant old world was being disrupted by the shiny, meritocratic new. It was a comforting myth, and one I embraced because I had seen comfort, without realising it was only a myth.

So for a while, I had been able to ignore the hints that the tech industry was, at the end of the day, just another industry. Even the clear indications of Silicon Valley hubris — the stories of corporate fraud or workplace harassment or overfunded startups — could be dismissed as exceptions.[5] None of them had been enough to dislodge my faith in the industry as a whole.

But the less I cared about my own startup, the harder it became to ignore the mounting evidence of the deeper moral rot within the industry. The burgeoning Me Too movement was spreading to the Valley, and the stories I was reading were convincing enough to overcome my knee-jerk tendencies toward victim-blaming. I watched as investors and entrepreneurs I had previously revered were ousted — only to rise to another position of power in due time. And the startups I used to admire were going out of business, their business models revealed to be built on unsustainable margins, worker misclassification, or customer deception.

I had always implicitly assumed that there was a core kernel of goodness within the industry, and so any problems that might exist would correct themselves eventually. The people within it were surely better than in other industries; startups grew the pie, so it was OK if their shareholders took a big slice of it; founders were driven by a genuine desire to make the world a better place — their philanthropy was proof of this. Everything I had been doing for the last two years rested on an utmost faith in the legitimacy of the industry. To admit that I had been wrong would be unbearable.

A year before, the *Wall Street Journal* had published the inaugural reporting on the numerous problems within Theranos, then valued at $9 billion. As more allegations of mismanagement trickled out, Liam, Nick and I had discussed them with an air of bafflement, trying and failing to make sense of them. We had rationalised their use of other companies' machines when theirs didn't work: that wasn't dishonesty, that was just good strategy! Everybody did it. Fake it till you make it, and all that. Ask for forgiveness, not permission. I had been so sure that

Theranos was the underdog, the victim of an unjustified attack by the media — they had raised almost a billion dollars! All that money couldn't have been wrong.

So when the whole thing came crashing down, I had found myself unable to make sense of it. I couldn't reconcile the company's credentialed board members and the prestigious investors with the increasingly incontrovertible proof of fraud. It didn't square with my understanding of the world. Accepting this one refutation of my premises would open the door to questioning all of them, and then what would I have left?

Now, though, in my post-Trump malaise, I was willing to look deeper than the post-money valuations, and so I was starting to see the industry in a new light. I learned about the workplace conditions of the forgotten workers who rarely got to share in the tech industry's success: the Uber and Lyft drivers making less than minimum wage,[6] the Amazon warehouse workers forced to work at a relentless pace,[7] the Foxconn workers whose low wages and brutal conditions enabled Apple's high profit margins.[8] Meanwhile, shareholders of all kinds rejoiced because their shares kept going up in value, and the newly minted tech billionaires were lavishly praised for pledging to give their money away[9] — even though they were still keeping enough to live opulently, and even though they were only giving to causes that accorded with their personal vision of what the world needed.

I had believed that the tech industry was different. I had thought that it pursued change for the sake of social good rather than the financial benefit of a measly few. Now, I was starting to see the stark divide between the few on top and everybody else, and it kind of looked like those on top were putting their own interests first — just as they did in other industries. I needed to find an explanation, and I didn't think I could find one among the VC blog posts and self-congratulatory founder stories.

In my disillusionment, I turned to something I had neglected for a while: books. I had been an avid reader when I was younger, but once I had started coding that had fallen by the wayside, with my book collection mostly limited to startup how-tos and Neal Stephenson novels.

By the end of 2016, our business was winding down and I didn't have much actual work to do anymore, so I started spending time at the local library. I wasn't really sure what I wanted to read, but I solicited recommendations from better-read friends, and sometimes I would just grab promising-looking books off the shelves. I was going in all sorts of directions: fiction, philosophy, economics, literary criticism, political theory. A lot of it went over my head, but every once in a while there would be a golden sentence ringing clear as a bell, and I would know in my bones that I was on the right track.

And there came an unexpected point when the strands converged. One strand started with popular accounts of the financial crisis which paved the way to deeper critiques of capitalism. At the same time, I dove deep on one particular American writer — David Foster Wallace — and, hungering to better understand his fiction, started reading literary criticism of his work. It was an essay about one of his novels that defined neoliberalism in a way that finally clicked for me, and I discovered that the term actually had meaning — it wasn't just a rude epithet hurled at Hillary Clinton supporters.

That was the first time it hit me that the books I was reading weren't merely a distraction from the real world. I'd been yearning for a deeper understanding of the tech industry and the world at large — maybe this was how I would find it. All these books had been under my nose this whole time, collectively representing a decent slice of the total intellectual output of humanity, but I had thought myself too good for them because I was a STEM major.

In late November, it occurred to me that I could study this stuff full-time. Liam's Canadian visa was running out, and he wanted to move back to London; I could move with him and do some sort of social science master's degree. I found the websites of all the major universities in London and scanned through their list of master's degrees, crossing out the ones that required a relevant undergraduate degree. One program, created fairly recently, described itself as the social scientific study of inequality. I wasn't sure what that meant, but it sounded intriguing. I bookmarked the application page and added the program's recommended books to my reading list.

In January, I started a personal reading challenge to up the stakes: I would read four new books every week. One book early in this challenge especially resonated with me. Immanuel Wallerstein's *Historical Capitalism*[10] — an unplanned library haul that happened to be on the same shelf as a recommended book — had a shockingly critical analysis of meritocracy. I had previously assumed that meritocracy was a concept invented in Silicon Valley in the 2000s, but this book was first published in 1983, and I had never read anything so clear and so convincing.

I used to take for granted that the tech industry was a meritocracy, and moreover that meritocracy was unambiguously good. A few years ago, one of my favourite tech companies had unveiled a rug on their floor which said "The United Meritocracy of GitHub", and when that had triggered a backlash from feminists in the industry, I had been baffled; my knee-jerk response had been to assume that the critics simply wanted the bar to be lowered for themselves.[11] More recently, having come to terms with how "meritocracy" was commonly used as a shield to deflect reasonable accusations of prejudice, I had become more cynical of the term, but without a good reason why.

But now I was reading a book written by a sociologist on a topic wholly unrelated to the tech industry that captured in words something I'd been groping towards for years, and it was simultaneously gratifying and mortifying to discover that people outside my field had managed to perfectly skewer it without even trying. I was building a more nuanced understanding of meritocracy: not only did the concept historically arise as a smokescreen for elites to justify their place on top in the face of emerging opposition from below,[12] but it would always be corrupted by the interests of those who got to define what constituted merit, and it would always be riven by the influence of other social inequities. At its best, meritocracy could be a horizon to aim for, a far-off utopia where everyone would be given equal amounts of opportunity; in practice, it was mostly used to excuse paying women less.

The startup trudged on. I continued sending out invoices and answering emails, but all my attention was now in the realm of books. I confessed to a friend that I'd been bored of my startup and had started reading anti-capitalist stuff to pass the time. The friend, a former philosophy major, suggested that I read Hegel, recommending a particular tome by Charles Taylor on the basis that Marx had built his work on Hegel's. I nodded as if I knew what that meant. Clearly I had more to learn.

In the meantime, we were still trying to get acquired. One potential acquirer was only interested in an acquihire, the sort where our product would be shut down and the purchase price would be converted into a stock grant with a three-year vesting schedule. It would be a good deal for the acquirer, as they would get new hires with a track record of working well together, and it would theoretically be a good deal for us, except that I was wary of committing to a full-time tech job for three whole years. But Nick had been fighting hard for this acquisition, and even though I had lost interest in this whole endeavour, I agreed to fly out to San Francisco for the technical interviews.

Our interviews were on separate days. The morning after we arrived at SFO, I walked over to their office in the Financial District. My day was packed with whiteboard interviews, all with engineers who evidently hadn't been briefed on the acquisition and accordingly treated me like a regular applicant. One asked me why I wanted to work here; I had not prepared for this question, so I awkwardly explained the context of the acquisition before mumbling something about the technical problems being similar to ones I'd worked on in the past. The interviewer seemed satisfied.

And then it was time for a lunch interview with some of their engineers. There was a surprising amount of gender and racial diversity among the interviewers, though I couldn't help but wonder if that had been stage-managed. I gave a lukewarm explanation of my startup, then asked them what their work was like. One engineer remarked that they got to work on really hard problems, then casually added that the smartest people he knew all worked here.

I almost dropped my self-serve taco, midbite. Did he really believe that? Wasn't that just what every company wanted its employees to think? And if he was right, what did this say about the social inefficiency of the system that put all these smart people to work on solving complex technical problems for the purpose of helping corporations sell more products?

In our early days, when I had been drunk on the Kool-Aid, I had believed that being smart was our main advantage.

Our competitors were dumb and slow, whereas we were agile and brilliant. It was as if the intelligence aspect was more important than what we were building — whatever we were working on had to be right *precisely because* we were smart.

But my startup remained small, with little funding and few customers. This was a billion-dollar company with offices around the world; they had HR, a Corporate Development department, a beautiful office in downtown San Francisco. They were doing the same thing we were, only they were a lot better at it, and now they were deeply integrated into the system. And that concerned me. I was starting to question the efficacy of the larger system which allowed this company to become so powerful, and which permitted its employees to believe that they were the good guys.

Interview over, I walked back to our hotel in Union Square. A couple of doormen were stationed outside to open doors for approaching guests. This puzzled me: why didn't the hotel just get automated doors? Surely that would be cheaper than hiring two people to stand outside and periodically open the doors. But of course, there could be other reasons to hire humans to do work that might appear easily automatable, and as much as I believed in automating away rote work, I had to acknowledge that the lack of automated doors created jobs for people who might not have had great options otherwise.

In any case, what did I think these doormen should have been doing instead? What right did I have to impose my value system of what I thought was inefficient on other human beings? Would I rather they be delivering food or filling boxes in an Amazon warehouse? Every second of their lives deprived of joy, squeezed out at the relentless pace of an unforgiving machine...

It finally hit me what I'd missed the whole time we

were thinking about automation: no matter how good our automation technology, there was no viable business model in which we would actually achieve our goal of giving workers back their time. We wouldn't be selling to the *workers*, after all; instead, we would sell our technology to the company, which would then implement it in a way that would allow workers to be fired. Rationally, the company would only buy our technology if it allowed them to get more value out of each worker — it wouldn't be financially prudent otherwise. As a result, the company would have higher profit margins; the workers would be downsized, or made to work harder. The workers were merely tools to be discarded as soon as they were no longer needed. Collateral damage was inevitable when you were trying to change the world.

It wasn't like we could sell directly to *workers*. If our technology actually worked, then the company would demand to buy it directly, and they would always be able to pay more than the workers. If we wanted to make more money, we would have to sell to management, which would be much more likely to put the cost savings toward enriching shareholders than to reward their workers. And who we sold to would determine our entire purpose, delimiting the possibilities available to us. We would have to make the product *for* management, and since the very essence of the product was to separate out the interests of workers from those of management, we would be prevented from acting in the interests of those workers.

I was starting to feel like the very act of private enterprise was in fact stacked against workers. If this was the case, then the economic system that glorified private enterprise must also be reliant on the structural subjugation of workers. It hit me that the clue was in the name — capitalism was literally designed to prioritise the rights of capital over the rights of workers. I couldn't believe it took me until I was nearly a quarter-century old to figure that out.

When we got back to Montreal, I immediately started working on my graduate school application, drafting a personal statement that linked together technology, capitalism and inequality. I was coming to the realisation that there was something wrong with the system itself, and I wasn't going to be happy merely carving out a place for myself within it.

A few days later, the company with the smart engineers got back to us: they wanted to acquihire us.

If you had told me three years ago that I would get acquihired by a San Francisco tech company, I would have seen that as success. Now, I couldn't imagine anything I'd want less than having to work for a tech company in that hellscape of inequality; the very idea of it nauseated me. I told Nick I had no interest in relocating to San Francisco, and he regretfully relayed to the company that we wouldn't be moving forward at this time.

Our other acquisition prospect — the company that had made us a tentative offer back when we had been too arrogant to consider it — was looking unlikelier by the day. We met with a founder of a company they'd previously acquired, who told us that his new employer was sobering up from a recent acquisition spree, and had frozen headcount. The company wasn't even profitable yet — they had only been able to make so many acquisitions because one of their deep-pocketed investors seemed to have an unwavering faith in the CEO.

If I had a choice, I would have much preferred an acquisition by this company over the San Francisco one — it gave off a more ethical vibe. But talking to this other founder disavowed me of that notion. His company's acquisition, for an undisclosed sum, had brought him and his co-founders to the US for prestigious jobs in management; all the non-founding employees stayed where they were, in a country famous for its cheap labour. As far as I could tell, their technology consisted of tools to outsource data validation and transcription to a low-wage country in order to conduct wage arbitrage. None of those low-level workers got to come to the US or own stock in their acquiring company, and their lives were probably little different. The founders, though, because they owned the company, were elated.

Of course, as a fellow founder, I was supposed to believe in the inherent virtue of founder-friendly acquisitions. The founders did the hard work to get their company started in the first place, and so they should be rewarded in recognition of their extra contributions, and as compensation for the risk they took on. But it seemed to me like that role was already overcompensated. Was it fair that founders got so much, when the employees whose labour kept things running got so little?

In February of 2017, there was a video going around Twitter featuring Nancy Pelosi, a US Senator for California and Speaker of the House of Representatives.[13] During a town hall hosted by CNN, a soft-spoken college student asked Pelosi whether the Democratic Party was planning to move left on economic issues, citing a poll showing that 51% of millennials supported socialism over capitalism. I was slowly connecting the dots between the concepts I was learning about in books and the divisions in contemporary politics, and even though I didn't entirely understand the question — what did it mean to move left? — I finally felt like I was digging in the right place.

The world moved on. Via social media updates, I watched

as founders I knew moved up in the industry: a successful acquisition; a massive series A; a pivot to a hotter space. I heard about former classmates getting prestigious jobs at fast-growing tech companies, including some who had previously embarked on very different career paths. Software really was eating the world.

There had been a time when I wanted nothing more than the trappings of tech industry success. I used to believe that tech money was *earned*, through a combination of hard work and intelligence, and I wanted the badge of having earned my achievements. If I had that, I wouldn't have to feel guilty, because the industry was doing great things for the world.

I thought about the founders who had unwittingly convinced me that I was on the right track. All these other seemingly smart people were staking their lives on the same startup dream — how could this path not be legitimate? Of course entrepreneurship was the most effective way to make an impact. Ignore the haters; just buckle down and work harder. Even if you hated your customers. Even if you had doubts about the utility of your mission. Even if you occasionally found yourself crying in the bathroom of your co-working space.

The way I had spent the previous two and a half years felt like a bad dream. At the ripe old age of twenty-two, I had plunged into a startup whose business model I barely understood in the sure-fire belief that I would find a pot of gold at the end of the rainbow. I had wanted to build something technically impressive and lucrative; I didn't consider whether it would be meaningful. I hadn't thought about it in terms of how I wanted to spend the next several years of my life.

I wanted to go back in time and shake my past self.

People had often told me I was brave for turning down Google in favour of starting my own company. The truth was, I hadn't turned down Google out of bravery, but out of fear — fear of something I didn't fully understand but knew I didn't want. And so I had sought refuge in startups, which I had thought could be a route to a loftier destination, far away from the banalities of the corporate world. A chance to start afresh, free from compromises.

I was finally starting to get it. The things that had turned me off Google, that had turned me off Wall Street for that matter, I had assumed were somehow specific, merely unfortunate aberrations in an otherwise solid system. I hadn't considered that startups might be subject to the same forces that governed the rest of the system.

When you started out, you were always the underdog. Everyone's expectations were low, so you had the freedom to explore and take risks. Your idea could be outlandish. You could have a delightful product that people loved without a revenue model. You could even give a chunk of stock to the artist who painted a mural. You were allowed to put fiscal considerations on the back burner.

What happened when you got bigger? Eventually, the profit motive kicked in; shareholders — public or increasingly private — needed you to think about their needs. Labour costs would be the first thing up for review. Stock options for non-essential personnel? Expensive perks? Health insurance and sick pay for roles where market research determined them unnecessary? Gone, if you even offered them to begin with. If your workers were compensated with piece rates — as was common in the gig economy — lower the price as much as the market could bear. You had to keep headcount low to appease Wall Street analysts carefully watching the profit-per-head figures, even if that meant overworked employees, impossible deadlines, and an army of frustrated contract workers lured in by the stillborn promise of a permanent role.

You had to protect your business interests, too. Get your software engineers to sign over all their intellectual property, even if it was developed outside the company, because otherwise an employee's side project might threaten your market dominance. This pile of ripe customer data? Don't waste it — find a way to monetise it, through advertising or flat-out selling it. The US military offered a contract that ensured your employees — many of them not even American — would be helping to fuel America's war machine? Your board wouldn't let you say no even if you wanted to.

That wasn't to say that all startups were good and all big corporations were bad. Plenty of startups were get-rich-quick schemes run by those who saw the industry more as a cash cow than as a way to enact meaningful change. It was just that startups had an easier time *appearing* to be good, whereas big corporations had less leeway because more people were dependent on them. But the point was that the startup stage was *temporary*. It was the larval stage of the large corporation, with all its foibles, all its historical baggage, all the ways corporations could mess up: underpaying workers, overcharging customers, engaging in socially detrimental practices for the purpose of profit.

Could there be another way to do it? Could there be a way to develop technology *without* the role of a multinational corporation necessitating patents and lawyers and stock price analysts? Could there be a way to build software that people wanted, without having to resort to business models that subjected users to endless distracting advertisements they would rather not see? Could there be a way to develop technology as a public good, shorn of the idea that it must make money in order for investors to recoup their investment?

Now that I was thinking about it, yes, that sounded great. I would love the freedom to work on something that I could wholeheartedly believe in, with the reward being not a huge payout but rather an average standard of living and the joy of being part of a society that catered to the collective good rather than individual greed. Of course, that seemed like a far cry from the world we lived in, and I had a feeling that getting there would require larger changes than me personally deciding to take a different career path. Focusing on individual optimisation just wasn't going to cut it for me anymore.

How was I supposed to unlearn a lifetime of meritocratic conditioning? For most of my life, I'd been driven by an individualistic determination to succeed. The terms of that success were defined by capital and refracted through the peculiar milieu of hacker culture I'd absorbed as a kid; but of course I thought them natural, rational, universal. At the peak of my ambitions, I wanted a prestigious tech job that would send me to conferences; I wanted a penthouse apartment in SoMa; I wanted a Tesla. And even when I stopped believing in the Google version of success, I didn't stop wanting *some* version of tech industry success. I wanted all my work to congeal into a larger narrative, the troughs of sorrow merely obstacles to laugh at from my triumphant pinnacle.

Sometimes the things we wanted were bad for us, whether we knew it or not.

What was I not seeing as a result of the path I had found myself on? What irrational beliefs did I hold as a result of my frankly absurd drive to be successful at all costs?

When the gig economy had first started taking over the streets, I had cheered it on. The phenomenon was clearly good for people like me — people who owned smartphones and could get a lucrative job working at one of these companies. Suddenly I could get food delivered, get a ride, or get my house cleaned at the touch of a button. Not only did I want this, but in some deep sense, I felt like I deserved it. And yet, in the absence of technology automating away every manual job, the necessary complement of these wants was that a whole class of people would be subjected to the whims of those with more disposable income. I had never considered whether it was reasonable for society to prioritise the consumerist desires of a few people on top, even if I was one of them.

If I wanted to understand this better, it wouldn't be enough to reason from first principles: my position in the system made me biased, and there were some things I couldn't expect to figure out on my own. I would have to seek out other perspectives as well.

In March, one of our partners finally agreed to buy some of our data, and we got a small cheque that allowed us to stay alive for a bit longer. Some of their employees had been frustrated with our lack of responsiveness in dealing with their complaints, and I didn't know how to explain to them that I had given up on the partnership, my startup, capitalism.

The most horrifying thing about all this was that it had all been for the sake of *advertising*. In the best-case scenario, we had been contributing to an environment of already pervasive advertising, where corporate logos trailed you wherever you went, products you looked at once followed you around the Internet for days, and brands sent you daily emails begging you to buy just one more pair of pants. That was the best-case scenario.

Our goal had been to help brands better understand

their customers so that they could better market to them. Had I really thought that the world needed smarter marketing? I thought it needed *less* marketing. I no longer wanted to build Tinder for advertising, and I didn't want to live in a world where such a product existed at all. Instead, I wanted to live in a world that had been liberated from the heavy smog of capital, where fewer surfaces were colonised by the output of marketing departments, and where cultural products could be funded without relying on the false generosity of multinational corporations with billion-dollar ad budgets.[14]

I had almost forgotten about my grad school application, and it was a pleasant surprise when I learned of my acceptance to the only place I had applied to. By then, my interest in the tech industry was virtually nil; I accepted the offer and started the process of applying for a UK visa.

For a long time, I couldn't muster the courage to give up on my startup even when I knew I didn't want to do it anymore. Doing a startup had felt a bit like engaging in my own personal war: everything was permissible except admitting defeat. I had been terrified of the idea of failure, especially the heavier kind of failure that required admitting you were wrong about your whole worldview. But that dark cloud was finally lifting; I could let go of my bygone hopes and look for something new. Unlearning past beliefs would not be easy, but maybe that was just part of growing older. Eventually, you had to shed the comforting myths of the past in order to face up to the bare desert of reality.

Nick returned to university to finish his degree. Liam and I packed up to move to England, in the process uncovering a trove of T-shirts acquired at various hackathons, conferences, and office tours. Some of the companies represented were long out of business; some were now worth billions of dollars. A gaudy quilt made out of these T-shirts would keep me warm in a Québécois winter.

If this was war, then I surrendered. I was out of ammunition, and the cause wasn't worth it anyway.

NINE: PROFILE BEFORE YOU OPTIMISE

The miniLab is the most important thing humanity has ever built. If you don't believe this is the case, you should leave now.
— Elizabeth Holmes, founder of Theranos, at an office party in 2011[1]

There was a programming maxim when it came to making improvements in a codebase: profile before you optimise. In non-programming terms: check the map before you head out. Don't just dive into whatever you come across; instead, start by running a profiler, which was a separate program that did diagnostics. The profiler would give you a better sense of which parts needed to be optimised — which parts you should prioritise.

The results could be quite surprising. The way a program actually executed on a machine could be quite different from your mental model of it. The parts with obvious inefficiencies that would be easy to optimise might in fact turn out to have little overall impact, whereas a section you had identified as good enough might be more inefficient than you had thought. If you sincerely wished to optimise your program, you would have to look beyond the low-hanging fruit.

In the past, I had believed that it was enough for me to work on an interesting technical problem. It didn't really matter what the technology actually *did*, as long as I was learning something and getting paid. I could have been tracking down dissidents for repressive governments, or algorithmically stealing tips from underpaid gig workers,[2] or trading people's personal data for the purpose of ad targeting — it was all the same to me. As long as someone was offering me a lot of money to work on a challenging problem, then on some level it must have been OK. I'd implicitly assumed that those with money could be trusted to spend it on the right causes.

But after nearly three years of building a startup that turned out to be nothing more than an interesting technical problem, I stopped believing that the technical problem was the most important thing; I wanted a mission that was actually socially valuable. After this degree, I wanted to go back into the tech industry, but this time I wanted to profile before I optimised. That meant I would need a deeper analysis of the way the world worked — a sense of the systems in place, the lines of forces, the blockages. A map, essentially, to navigate Silicon Valley's euphemisms and obfuscations and flat-out lies. After all, everybody wanted to believe they were the good guys, and I would need the ability to distinguish between what was real and what was marketing.

Soon after arriving in London, I attended an academic conference on artificial intelligence that I had heard about through social media. I was mostly here to observe; it had been ages since I last went to one of these conferences and I kind of missed it. Plus, maybe I would find some answers to the questions that had been nagging me of late.

It was July 2017, and the techlash was burgeoning; ordinary people were worried about AI coming for their jobs and unscrupulous tech companies misusing their data. The European Union had recently adopted a wide-ranging privacy regulation called the General Data Protection Regulation (GDPR), and companies had less than a year left to enact it. Formerly vaunted tech companies were making the headlines for all the wrong reasons: their byzantine schemes for tax avoidance, fines for careless handling of data, exploitative labour practices. I suspected that there would be more backlash to come, and I wanted to know what academia was saying about it.

The conference was primarily technical, focused on the mechanics of different systems for performing computational learning. But, perhaps as a nod to the growing interest in the field's social responsibilities — concerns over gender and racial biases encoded in data, for example — the conference had convened a panel on how data regulation would affect the field. Key to the agenda was GDPR's "explainability" clause, a little-known but potentially quite expansive requirement that decisions made using machine learning be "explainable".

As the all-white, all-male panel discussed the technical feasibility of this achievement — with the requisite dunks on regulators for drafting something so technically ignorant — I furiously jotted down notes, biting down the urge to interrupt. Some of the panellists seemed to think that GDPR was too radical, for it would force companies to disclose information that would be difficult to synthesise and might not even be relevant. From my perspective, though, GDPR's mandated explainability clause was not even close to radical *enough*.

The value of explainability was to help people understand why an algorithm made the decision that it did. The more complex the algorithm or the data set used, the harder it would be to explain in a way that made sense to humans, who were rarely good at envisioning hyperplanes in n-dimensional space. But you could still devise heuristics to sufficiently approximate the explanation, even if the system was well beyond the understanding of any mere mortal. If a predictive policing algorithm led to a black person being pulled over, or a job candidate got rejected by a hiring algorithm because her resume listed a women's college, the involvement of the algorithm would not exculpate the actors. If the algorithm was trained on real-world data, then it would of course recycle the same historical biases that existed in the past.

What would explainability add here? You could imagine a little ! icon next to the automated decision statement triggering a popover explaining the factors behind the ruling. Great, here were the algorithm's biases. Now what?

Explainability was the *first step*, not an end in itself. As much as I was intrigued by the technical challenges involved, it struck me that the upshot of explainability was to uncover existing societal inequalities hidden in the data. The hard part was what would come next, in the contestation. And I didn't know how that part was supposed to work.

I listened to the panel, straining for even a throwaway mention of the social problems that would come after explainability. But the panel's focus seemed much narrower: how to do the bare minimum to comply with regulations mandating greater explainability for consumers. Nothing at all about a theory of change.

Maybe I was missing something. Maybe everybody knew that once the biases were revealed, then other institutions would step in. Sunlight was the best disinfectant, and all that; people would file lawsuits, regulators would slap fines, engineers would proactively write code to counteract biases. The system would correct itself, with time.

And yet, we already *knew* that these historical inequities existed. It seemed to me like the right place to optimise wasn't the knowledge aspect, but rather the accountability part. For things to change, you needed more than people knowing that they were being discriminated against; you needed the powerless to build power. And that was very much not a technical problem.

My masters program wouldn't start until the fall. For the rest of the summer, with my startup basically defunct, I plunged into books with a hitherto unseen enthusiasm. My book-reading pace neared one per day, still without much of a discernible pattern or strategy. I was mostly following the trail of breadcrumbs from other books I had read, all of which seemed to fit loosely within an unfamiliar part of the political spectrum which described itself as "the left".

I couldn't get enough of it. It was like discovering a new world. I learned that Universal Basic Income had not actually been invented in Silicon Valley[3] — it was an idea with a long and contested history, and it could be used for vastly different political ends depending on the terms of its implementation. I learned that the gig economy was merely a technological twist on a very old pattern of capital attempting to pay workers less.[4] Despite not having a formal program, or even a firm sense of what I was looking for, with each pure note I was solving an unspoken mental puzzle.

In learning these concepts, I found an unexpected parallel to when I first began a formal education in computer science. Before university, I had done enough programming on my own to have an intuitive understanding of various concepts, but I had no sense of how things worked behind the scenes, nor did I always know the accepted terminology. I learned through trial and error, substituting random symbols in a regular expression or variables in a function to see what would happen — a painstakingly self-taught journey of discovery. It was only in university that I finally learned the words and explanations for things I'd previously accepted as mysteries.

And so whenever I would first encounter the formal definition for something I had previously understood only by feel, there would be a bolt of recognition as the new concept elegantly slotted into place. It would be like opening a whole new dimension of understanding — like the difference between watching a dance, and dancing the dance. Everything would just click. It wasn't so much a case of learning something new as it was of putting words to something I'd already intuited, allowing me to assemble a skeleton on which to drape my understanding of the world. An analytical framework was coalescing in my mind, and I was starting to comprehend what I had once thought would always be incomprehensible.

With the zeal of a new convert, I scrutinised the catalogues of various left book publishers and subscribed to a bevy of periodicals. Not all of this analysis was meaningful to me, but I was captivated by the potential of this new paradigm and I wanted to soak up every last drop of knowledge. Everything looked different in this light, and I couldn't believe I'd been in the dark for so long.

I was beginning to understand what I had missed. My erstwhile faith in the tech industry had been rooted in an assumption that technology would liberate people, giving them back their time and increasing their control over the world. After all, that's what it had done for me; it gave me a community, the heady feeling of power over a machine, the possibility of a lucrative career doing something I profoundly enjoyed. But technology could have another effect: that of augmenting the power of corporations, in a way that might be orthogonal to actually liberating people. I had never considered that, for some, technological change might just be another painful thing to endure.

Now that I was primed to look for it, I saw signs of the left everywhere. In London, I found radical bookstores and book fairs filled with books I'd read and wanted to read; in Berlin, I walked past G20 protest posters featuring slogans about fighting capitalism. I suddenly wanted to revisit all the important events that happened in my lifetime, to see if there was a side I had missed at the time. Occupy Wall Street, the red square protests at my university — my instinct had always been to side with the status quo, under the assumption that the anti-systemic protests were juvenile because they lacked clear demands. I had never considered that it might in fact be reasonable to reject an entire system.

In London, I showed up to a barbecue that had something to do with left politics. I didn't know a single person there; I had found the event through a retweet by someone I followed on Twitter. But the people I met were welcoming; we had read the same books and had the same criticism of established institutions, and I was thrilled to finally meet people I could talk to about these things.

One person I met worked part-time for a gig economy platform. His real passion was creative, but it was a highly insecure profession, and he did delivery work between auditions to pay the bills. This was my first time meeting, in a social setting, someone working on the *other* side of the gig economy. The side that someone like me would expect to join came with a decent salary, free food, stock options; on the other side, there was no way to negotiate pay — you had to accept whatever rates you were offered, hoping that it would be enough to cover your expenses. You weren't even allowed to refer to the company as your employer, because their business model necessitated employee misclassification. And if you ever got injured on the job, they were more likely to hound you to complete your task than to look after your wellbeing.[5]

I thought about my own fortunate journey — all the contingencies that could have easily gone the other way. What if my passions had led me to a labour market with a less favourable balance of supply and demand? What if I had been less obstinate about persisting as a programmer despite the myriad signs that the industry was not always welcoming to someone like me? What if the people who had interviewed me for my first job had decided not to take a chance on me?

So much of my self-worth was built on top of the idea that I deserved what I got because I worked hard. The flip side of this belief was that those who were less successful were less deserving, precisely because they didn't work hard enough. And yet it would be naive to think that meritocracy worked for everybody just because it worked for me. Not everybody had the opportunity to climb the ladder, and in any case the terms of that climb were arbitrary, based on what was most advantageous to the people in charge. How appalling to find self-assurance through the perceived inferiority of others.

Late September, my master's program began, but the dry lectures already seemed dull compared to everything I was discovering outside academia. One weekend, I took the train to Brighton for a multi-day festival about left politics, where I listened to panels featuring speakers whose books I had devoured.[6] They shared stories of struggles that were going on now — efforts to resist the worst excesses of late capitalism — and for the first time in a while I felt hopeful.

It was strange to suddenly meet so many people outside the tech bubble. I was embarrassed to explain my startup melodrama to people who were working multiple jobs yet still drowning in student debt that they suspected would follow them to the grave — my acquisition woes sounded even more frivolous in comparison. Most of my tech friends didn't have student debt anymore, if they ever did; after a couple of years in the industry, they would amass enough stock grants to pay it off. Even a high-end debt of several hundred thousand dollars would be manageable if your company were to IPO or get acquired by Google. But if you were a schoolteacher or nurse or gig worker, there wouldn't be any looming IPO for you to look forward to; Google was not about to acquihire you and pay off your debts. This seemed like an altogether worse system than not having debt at all, in a world where education was treated as a social good rather than an individualised investment in one's earning potential.

By October, I learned that I wasn't the only tech worker suddenly discovering the urgency of progressive politics. *Tech Against Trump*, a short book published by *Logic* magazine in 2017, featured stories from fellow tech workers who were startled into caring about politics because of Trump. I discovered the possibility of political action beyond the realm of electoral politics, through building workplace power within a well-trod framework of workers' rights, and I realised the power of seeing tech workers as *workers*, with all the political implications associated with the term; even if their job seemed more glamorous than being a coal miner or warehouse worker, they still had grievances which could be addressed through building collective power.

Previously, I had only had a hazy understanding of what a trade union was. In my mind they were a bad thing: a harbinger of inefficiency; a threat to the growth of beloved tech companies like Uber and Lyft; somehow related to organised crime. I had seen Paul Graham's tweet that the presence of unions implied potential energy that could be unleashed by startups,[7] which I had thought was reasonable at the time — unions artificially inflated workers' salaries, right? They distorted labour markets, whereas startups could disrupt them and restore efficiency.

Now I was seeing the other side of the analysis: that some people were workers because they had nothing to sell but their labour. There was no neutral analysis of the "right" salary someone should be paid; there was only relative bargaining power, and a skewed balance of power would spring up all sorts of excuses to justify that imbalance. And capital, in its strength, had no compunction about taking everything it could from its workers in exchange for the means of bare existence. Workers who unionised were doing the only thing they could to protect themselves from a much more powerful force, because capital's prerogative was to treat them as worthless in order to extract as much profit as possible. Paul Graham's perspective on unions might be less a dash of universal wisdom and more a reflection of his personal position in the economic system.

In November, I attended a Marxist conference in London.[8] One session was a discussion of technology in the workplace, and the speakers shared war stories of sabotage as a means of resistance: intentional slowdowns to keep the algorithm from inflating the chosen targets, deliberately messing up tasks to buy time. Previously I couldn't have imagined that a worker would ever want to sabotage their company, but now I was starting to get it. Not everybody had the luxury of working a job they liked; sometimes the only choices were lacklustre roles where you were told to perform rote tasks by an algorithm, never reaping the benefits of increased productivity, the hours burnt up in order to pay the rent. The urge toward sabotage was no less rational than compliance, because at least it reclaimed some agency in a world that offered most workers very little.

One speaker talked about his work delivering food in the gig economy. He usually worked in the evenings, right after a long shift at his other job; after six hours of biking around the city, he was typically exhausted. Afterward, I asked a question that made sense to me from my newly tech-hating vantage point, but which evidently struck the speaker as naive: given that the work was so awful, shouldn't we just get rid of the gig economy? The speaker politely disagreed, saying that gig work was still better than any other job he had ever had; he just wished it paid better.

And just like that a massive puzzle piece clicked into

place. The distaste I had for tech companies, especially gig economy companies, was a little misplaced — it wasn't just about the tech industry. The tech angle might explain why these companies were so well-funded, but that was incidental to the deeper problem, which was that workers *didn't have enough power*. Otherwise, they could demand better wages, healthcare, benefits; they could push for a better social contract so that delivering food for six hours straight every night was not something anyone needed to do merely in order to pay the bills. What's more, they could ensure that that their work was put to better uses than it currently was, according to the logic of the market.

I thought about the software engineers building the technology that dispossessed other workers of their agency. I couldn't really blame them: if your employer continually insisted to legislators that the Dashers or Shoppers or Associates who did all the manual work were users rather than workers, it would be understandable to not see them as your fellow workers. And if your job was to build the app they interacted with according to metrics like customer satisfaction or order volume, it would be only natural to start seeing workers as tools rather than human beings. Like Amazon Web Services' EC2 servers: spun up when needed, shut down when not. And it would be convenient for your company's bottom line to pay these workers only for the time they spent working. If that were to result in workers orchestrating their lives around the black box you'd built — and suffering as a result — that wouldn't be your problem. You were just doing your job.

If you were a software engineer with a decent track record, you could generally expect to move up the career ladder over time, watching your net worth go up and up. Upward mobility — while not guaranteed, and not available evenly — would be a reasonable expectation for most. If your startup were to shut down, you would probably be able to find a job with similarly lavish pay and benefits. And if your startup were to hit the lottery, you would have a solid chance of becoming unimaginably rich.

But if you were an independent contractor working through a gig platform, the paths upward would be much murkier. You could work one-hundred-hour weeks and still barely pay the bills, much less build up your savings. Job security would be an illusion; your pay could drop at a moment's notice, and you would have little recourse other than to accept it if you didn't want to lose your work history on the platform. And if your account were to get deactivated, your only hope would be to talk to a customer support agent who would probably also be a contractor with little power.[9]

These were two fundamentally different notions of work, one of which seemed a lot more punitive than the other.

And I wondered: how long would engineers be safe? At the moment, tech companies were flush enough that they could afford to pay well for important roles, and there was an undersupply of candidates with the necessary skills to jump through recruiting hoops; job prospects varied by market and industry, but generally they were quite good. But the imbalance of supply and demand wasn't going to last forever, and in fact many tech players were already investing in ways to increase the future supply of engineers.[10] What would happen when funding dried up, or tech companies realised they could get away with lower offers? Who would manage to cling to their privileges, and who would get the axe?

It certainly wouldn't be the workers themselves who would get to make these calls. They might have some autonomy over the code they write, but they generally did not have autonomy over management decisions; they would ultimately be hemmed in by the same forces

acquisition, but I had given up caring; the part of me that equated money with value was dead. I just wanted to get it over with so I could move on with my life. I said a last goodbye to the codebase, which for a while I had seen as my crowning achievement; I was still proud of what I had built, but I also now knew that it didn't matter. I transferred the repository over to the new owners, deleted the outstanding items from my to-do list, and silenced the part of my brain that still occasionally looked at people and saw data. Not my job anymore.

In the meantime I was sprinting away from the tech industry and toward a whole new world. Prior to my degree, I had devised a hazy plan to get a job at Google in some sort of policy or ethics role, but now I was convinced that any such role would ultimately serve as corporate window-dressing with little power to actually effect change. I read about labour law and the history of worker activism; I attended protests and joined my first picket line; I started a weekly seminar group on radical politics with speakers ranging from union organisers to cultural studies scholars to left-wing policy makers. I wrote a first-person essay about my disillusionment with the tech industry[13], in which I suggested that tech workers should see themselves as workers; it spread like wildfire, and I received emails from people all over the world telling me how much it resonated with them.

From across the ocean I watched the surprising wave of tech worker resistance in the heart of Silicon Valley. Tech workers were organising around the identity of worker in a way that I could never have imagined before. Their companies might be worth billions of dollars, but all that wealth wasn't trickling down to them, and they were carving out an identity separate from that of the people who called the shots. One software company, Lanetix, saw its software engineers form a union and summarily fired them with plans to ship those jobs offshore; software engineers from other companies joined their demonstrations.[14] Microsoft employees were protesting their company's contract with Immigrations and Customs Enforcement (ICE) in the wake of recent revelations about the brutal conditions at the US-Mexico border.[15]

I was especially fascinated to watch what was happening at Google. Google workers were among the most militant in agitating against their employer: a contract with the US military had become the focus of a company-wide uproar,[16] and in May, multiple engineers working on the project resigned from the company to cement their refusal. And in July, *Bloomberg* featured a story about the company's contract workers now outnumbering full-time employees, which they called a "shadow workforce"; the way they described it, it felt like a caste system, with one of the richest companies in the world treating most of its employers as second-class citizens for the purpose of making their financials look better.[17] When I had worked there, I hadn't understood the reasons behind the contractor/employee distinction; I had intuited that the IRS was somehow to blame for contractors not being allowed at company parties.

There had been a time when teachers hadn't thought of themselves as workers, and consequently shied away from efforts to organise collectively. But now, a teachers' strike wave was spreading across the US, and they were winning concessions they wouldn't have gotten otherwise.[18] Maybe tech's time would come, too.

Late October 2018, I happened to be visiting San Francisco when Google employees announced a mass walkout worldwide.[19] It was scheduled to take place on November 1 at 11:11am, and it was in protest of Google's handling of a sexual harassment investigation, which had left the harasser with a severance payment of $90m.[20] Women at the company were especially furious at the news, and even Google loyalists were awakening to the realisation that their seemingly progressive company preferred to protect the powerful over low-level employees.

I was staying in SoMa, and I took a Lyft up to the Spear Street office I hadn't seen in half a decade. An old friend was working out of the office that day, and when I saw him on the sidewalk, he explained that he wasn't particularly moved by the walkout demands but had decided to join anyway because everybody else seemed to be here.

The sun was shining as the marchers filled the sidewalks and spilled out onto the mostly empty streets. There were hundreds of Google employees there that day, carrying signs about womens' rights and workplace democracy, and I was a little giddy as the delegation started marching towards the Embarcadero. At Harry Bridges plaza, the organisers gave short speeches about workers' rights and read out horror stories from anonymous workers about their experiences with Google HR. People cheered at the right moments, though there was a faint air of uncertainty as many of the attendees looked like they were unsure about the right etiquette for a protest, or whatever this was; some clung to their phones looking like they would rather be back at their desks. Baby steps, I supposed. Getting a bunch of white-collar tech workers to come out for an action that was specifically in protest of management — to acknowledge that the company was not a family, but rather consisted of distinct classes with distinct interests — was enough of a win for today.

When the speeches ended, people started streaming back to Spear Street. On the walk back to where I was staying, I passed a picket line outside a Marriott hotel. Marriott workers had been on strike for almost a month, as part of a contract negotiation between management and the union, and workers were holding signs that said "One job should be enough". Most of the workers on the picket line were Chinese women. When I grabbed a resting sign and joined the line, a few of the workers smiled at me. One worker asked in Mandarin which hotel I worked at, and when I explained that I was just here as a supporter, she insisted on giving me some of their noodles.

On the surface, you would think Google engineers and Marriott hotel cleaners couldn't be more different. And yet, one key component of the hotel workers' union dispute was the prevalence of sexual harassment in the workplace: when male guests harassed the mostly female cleaners, management would often turn a blind eye, and cleaners would sometimes even have to return to the same room.[21] The specifics might be different, but the same underlying problems existed at both companies. Google workers might currently experience certain advantages over service workers, but at the end of the day, they were still working for a corporation that would put management's needs ahead of workers'. When it came to class interest, Google and Marriott workers would have more in common with each other than they would with their respective management. One struggle, one fight.

For a while, I found it difficult to talk to people in the tech industry, even old friends, about my massive personal pivot. In all honesty, I was still trying to make sense of it myself. Plus, I was hesitant to say too much because I was worried about offending them — or boring them. But the people I talked to were generally fascinated, and I found that a lot of people I assumed to be happy with their careers were in similar states of dissatisfaction to my own. Some had mentally checked out; some were quietly mulling over the alternatives or simply biding time until they'd saved up enough to quit. It was a bit of a weight lifted off my shoulders to know that I wasn't alone, and I was grateful that I had the ability to quit when I did.

Sometimes I would get nostalgic about my old life. Things were easier back when I had structure, a consistent path to follow, simple metrics to optimise for. Once in a while I would check the website of my accelerator program to see the logos of startups whose founders I had gotten drunk with, whose pitches I would hear over and over in the office. Some had shut down or gotten acquired; some had undertaken bizarre pivots. I wondered how many of them were merely stumbling on as somnambulistic shells because they were too scared to admit failure — like we were, for so long.

I supposed any portfolio of startups would eventually become a mausoleum. That was just how it worked: lots of failures in exchange for a small number of successes. It was a nice system for venture capitalists, who got to swan around in their Teslas and Allbirds and Patagonia vests; they got expense accounts, carried interest, the thrill of being worshipped by hopeful would-be entrepreneurs. What was it like for founders? Your investors could choose to give you some degree of autonomy, but ultimately it was *their* choice; your life's work was not truly your own. All that time you could never get back, spent in the pursuit of a foolish cause, governed by forces out of your control. Your hopes and dreams over-determined by a system requiring you to sacrifice your time for the goal of getting your investors a return on their investment.

Not to say that founders deserved more sympathy — even the failed entrepreneurs would probably be fine; the industry was tolerant of failure to a fault. But it did feel a little pointless all the same. All the exertion, the sleepless nights, the incessant disruption— for what?

TEN: STUPID ENVIRONMENT

Computing is terrible. People think — falsely — that there's been something like Darwinian processes generating the present. So therefore what we have now must be better than anything that had been done before. And they don't realize that Darwinian processes, as any biologist will tell you, have nothing to do with optimization. They have to do with fitness. If you have a stupid environment you are going to get a stupid fit.
— Alan Kay, a computer scientist who contributed key work to object-oriented programming and graphical user interfaces[1]

Consider Silicon Valley as the end state of a fitness function.

This function is guided by the accumulation of capital: what makes money is what gets funded. Of course, it's not entirely deterministic — there's room for different approaches, different degrees of profit versus risk — and it's not totalising. Some money will flow towards causes that aren't profitable on their own, but which enable the extraction of profit elsewhere, and there's even room for initiatives that aren't aligned with the profit motive, though those will always struggle to expand as long as a capitalist resource allocation function dominates.

As tempting as it is to imagine technology as a refuge from the real world, technology is never developed in a vacuum; its funding depends on the larger political and economic environment which constitutes the fitness function for allocating resources. And for the last forty-odd years, much of the world has been gripped by a turn toward neoliberal policy,[2] with US monetary policy paving the way. The balance of power between capital and labour has decisively swung in favour of capital, leading to a world where decisions about how resources should be spent and what people should be doing are made for the benefit of large, often multinational corporations.

If labour ever had much of a say in economic matters, it certainly doesn't after the onset of neoliberalism. Labour unions are in steep decline; politics has been reduced to choosing between whichever capital-supporting party is the lesser evil; and most people have little control over the workings of even the tiniest sliver of the economy. That control belongs to capital, in its newly globalised, heavily financialised variant; it belongs to billionaires, and to those guiding policy on their behalf. It certainly doesn't belong to the workers — toiling in sweatshops in countries with lower labour standards, or working in the regulatory grey zones of the shadow economy — nor to democratically elected governments, which are often browbeaten into offering generous tax breaks under the banner of job creation.

In this neoliberal desert sprouts the seed of Silicon Valley. The mid-century semiconductor revolution soon gives way to the birth of the Internet and the personal computing devices that can access it. Within this desiccated wasteland that is the proverbial end of history, an arid capitalist and neo-colonialist realism is the order of the day. What's good for corporations is good for America, and what's good for America is good for the world. Technology might be initially funded by states, but it's always rolled out through commercial means. Corporations are just leaner, more efficient, *better* — especially those headquartered in the developed world.

By the end of the millennium, capital is effervescent, sloshing around the system thanks to a friendly regulatory environment, the rise of financial engineering, an intensification of natural resource extraction, and wage concessions exacted from a defeated labour movement. It then floods into the tech sector, buoying up valuations beyond what can be sustained even by a market this irrational. When the bubble bursts in the form of the dot-com crash, leaving behind job losses and half-baked promises in its wake, the money simply sidles over to more promising ventures, like inflating real estate assets.[3] And when that collapses in the form of the subprime mortgage crisis, the money slowly trickles back into tech, in thrall to a collective amnesia about the causes of the last bubble. That surely wasn't the moribund gasp of a diseased economic system — it was simply too early. *This* time is different; *this* time the tech industry will be worthy of its valuations.

And for a while, it seems like everything is fine. From the ashes of the first tech bubble rises the new and improved Silicon Valley: dot-com-era companies that survived, cockroach-like, and are now cash flow positive; retries at old business models within a more friendly funding environment; plus a whole array of new business models taking advantage of recent technological advancements. And these companies have no shortage of innovative business models. Some commercialise the fruits of state-funded research using the labour of faraway workers driven to the brink of suicide. Some come up with clever ways to sell people's attention at scale. Some sell products at a loss in a bid to capture the market by driving competitors out of business. The successful ones build solid moats in order to maintain control of their platforms, aggressively enforcing IP infractions to ensure that their technology remains theirs.

The bigger the industry gets, the more undiscriminating the money that flows in. The ventures become more outlandish and more brazen at flouting the conventions of generally accepted accounting principles. You don't need to be *profitable* to raise billions of dollars in this economy, as long as you have a plan to eventually control a sector of the economy in a way that convincingly reminds your investors of Amazon. You don't even need software margins to be valued as a tech company: you just need an elegant logo, overinflated executive compensation, and Kombucha on tap, wrapped up in a feel-good corporate mantra proclaimed in a tasteful sans-serif.

Is this just the morose fate of technology, to be a signifier for the most profitable outlets for capital? Are we doomed to remain in this bubble of irrational exuberance, with startup valuations perpetually soaring beyond the realm of known price/earnings ratios? In a sense, our current predicament is not irrational at all — it's highly rational under the prevailing economic logic of capitalism. The promise of future profits through clever applications of technology is almost always going to be a more prudent investment than non-profit endeavours to make people's lives better. This is why we now have a surfeit of consumer products catering primarily to rich people even as the region that many of these startups call home is teeming with people who have been made homeless.[4]

It is a stupid environment, and it has produced a stupid fit.

In a capitalist system, money is drawn toward the tech industry in the hopes of making a return. Some of this may be driven by altruistic motivations, out of a genuine belief that the tech industry is the best vehicle for creating a better world, but there's always an asterisk to the better world being created: a world where the investors make a healthy return. This is the heart of the chief economic imperative driving the tech industry, whose real product isn't smartphone apps, or even smartphones — its chief product is return on investment. And this is something that the industry has excelled at.

Even Silicon Valley believers can no longer ignore the broken ecosystem they've helped produce. Some suggest democratising funding at the level of venture capital, by backing more founders from historically underrepresented groups. And sure, this might be a step up from the current situation, but the disease is deeper than the composition of who gets funded. It stems from the very raison d'être of the industry — that ventures need to seek a return, and so the profit motive is a valid driving force.

Of course, defenders of the status quo may disagree that there's something fundamentally broken about the current system. They may acknowledge the myriad failures, the Theranoses and Juiceros and WeWorks, the overvalued and overextended companies that may never be profitable — but they would argue that those are bound to happen in any system, rather than being an indictment of the system's failings. They would insist that it's still worth it in the end. But what are the successes that justify the copious amounts of capital dumped into this industry? Do we need multiple billion-dollar scooter companies?[5] An endless parade of cryptocurrency companies, and companies selling virtual shovels to other cryptocurrency companies? The next hotel chain or car company or glorified landlord valued at tech company valuations because there's too much oil money floating around?[6]

And even the putative successes — the technology we love — are not ours. The products we cherish, the ones that store our data or memories or connections, are owned by shareholders and operated on their behalf. There is no democratic recourse when the company that makes a beloved product gets bought up by a bigger company and shut down. Users have few avenues to contest a company's decision to retroactively delete their data, or to start displaying ads, or to recommend certain content over others. And it doesn't matter how crucial a piece of software is to a particular ecosystem, or how much community work is involved in keeping it alive — it ultimately belongs to investors, and those investors have just installed a new CEO whose goal is to squeeze out some revenue.

That's not to say that these products shouldn't exist, or that they should only be built outside the tech industry. It's to say that the implicit profit motive that directs the industry's operation is a bad fit for technological development. And in the current frothy macroeconomic environment, valuations have been pumped up beyond reason, "good" and "bad" startups alike. Not only does this influx of capital poison the useful products by encouraging them to sell out, but it also inflates the valuations of less useful companies.

And here's the thing when startups are catapulted up to astronomical valuations by an inrush of global capital. Past a certain point, it's no longer a game — it's real economic resources. The absurd valuations cited breathlessly in the tech press are not just numbers out of thin air. Money represents an entitlement to a share of the social product, and the current tech bubble is effectively drawing away resources from more socially useful endeavours.

I keep coming back to this: it's all the same money. Each dollar of VC funding for another useless startup in an already crowded space is the same legal tender that allows or denies someone access to housing, food, insulin. The finite resources in this world — natural, land, people's time — are all managed through one system, and this system is currently over-indexed on encouraging corporations to pursue profit at the expense of societal welfare.

No individual actors in the ecosystem are to blame for this, of course. No one in the tech industry is directly responsible for its role in siphoning resources from more socially valuable causes. The system is bigger than them, and even if they mean well, so much of it is out of their control. The problem isn't merely the individuals who make decisions that end up hurting others; the problem is the system where ill intent is not required in order to do harm.

In the face of all this, it's tempting to believe that the industry is still ultimately benefiting the masses. A common argument in this vein is that profits from venture investments flow back to pension funds and university endowments, and so even if some companies seem of dubious social value, at least there's some external benefit to the inflated valuations.

It's a promising line of thought, but only until you break down the actual numbers. Venture capital firms' limited partners are not restricted to pension funds and university endowments — other sources of funds include the wealth stores of dynastic families (like the Waltons or the Kochs), sovereign wealth funds (often based on oil sales), and the cash hoards of corporations who have cleverly minimised their tax bill.[7] And the list of beneficiaries of VC-backed wealth creation includes founders and early employees who might receive astronomical windfalls, not to mention the partners at these investment firms who get a share of the profits. The full distribution of where the money goes is usually not public, but it's misleading to pretend that the *primary* beneficiaries of the tech bubble are pension funds and university endowments. Relying on those two is a diversion to distract from the less savoury recipients.

In any case, neither endowments nor pensions can be said to be an unabashed good. University endowments in their current form exist in response to a privatised landscape where education is seen as a commodity rather than a public good. The institutions with the largest endowments tend to be sites for reproducing privilege, with unreasonably high tuition costs, a high proportion of legacy admits, and a capitalist division of wages whereby high-up administrators draw millions even as lecturers, researchers, and service staff are undervalued and prevented from unionising[8]; meanwhile, universities that primarily serve lower-income students have few resources and little cultural standing. The current system optimises for sorting individuals into a hierarchical society, not equity or social good; if we wanted the latter, schools would be funded collectively through progressive taxation, rather than left to fend for themselves.

The merits of the present pension system are also questionable. In the US in particular, pension funds are tied to work and benefits are indexed to contribution, essentially making them a reward for having worked in certain jobs. This isn't the only way to ensure that people are able to retire; more social democratic countries treat pensions as a collective good, funded again through progressive taxation rather than individualised and subject to the whims of various overpaid investors. Pensions could be seen as one small component of the social contract, part of living in a society where we take care of other people. In any case, the minor role played by pension funds in venture capital does not make up for the harm caused by VC-backed companies. If some workers' pensions are currently tied up in the continued growth of an exploitative company whose labour practices endanger other workers, then this entanglement is a problem in itself, and the system needs to be redesigned without such perverse incentives.

In short: both the financialised pension system and the private university endowment system are suboptimal. They are at best local optima of the capitalist fitness function, and both need to be radically reformed. If they are the main justification for VC's continued existence, then VC's utility seems limited, only suited for a subpar environment where money is indiscriminately dumped into private enterprises under the assumption that the market will sort it out.

And what guarantees do we have that the market will optimise for the right ends? The logic of the market is to encourage self-interest in the belief that the invisible hand will make everything right, on the grounds that economic activity is a good in itself. But the measurements we use for the health of the economy — GDP growth, unemployment rates, stock market capitalisations — are not themselves infused with ethical values, and it's becoming increasingly clear that they are a poor proxy for tracking societal wellbeing.

Purdue Pharma earning billions from selling the drugs that fuelled the US opioid crisis?[9] That's entirely reasonable by market logic. Clothing manufacturers and food producers destroying excess goods in order to preserve the scarcity needed to maintain high prices? They have to protect the value of their brand, after all. Low-wage employers subjecting their workers to unnecessary stress and injury in order to shore up corporate profits?[10] This might actually be good for the economy — a whole bevy of enterprises has sprung up with commodities to salve their boredom, payday loans to financialise their poverty, for-profit healthcare to treat their wounds. When everything's a commodity, even human suffering can be an economic boon.

Ecological damage isn't necessarily a problem under market logic, either. What else can you conclude when fossil fuel companies are valued under the assumption that their reserves will be entirely depleted, even though that would cause untold physical harm to the natural world? And what about the ecological damage of globalised commerce, which thrives on the power imbalance between multinational corporations and local populations in resource-rich, low-wage areas in order to transport and then sell goods at a mark-up in wealthy markets?

The reasoning behind market logic seems to be that if people will pay for a product, then the product is good. This is myopic at best: at most it means that the product is useful in our current suboptimal environment. Of course customers will pay for VC-subsidised rideshares in cities with underfunded public transit; of course people will resort to payday loan apps when their boss doesn't pay them enough to survive. The ostensible success of private enterprise does not imply that private enterprise is the best way to address the underlying problems.

Because here's another way to understand profit: as an indicator of inefficiency, a sign that something has gone wrong in the factors that shape a given market. Some level of profit may be acceptable in the early stages, or when necessary for expansion, but if a corporation has swelled to billion-dollar heights and is still turning to new frontiers to maintain profit margins, that's a problem.[11] At that stage, profit should be attributed to some combination of these distasteful possibilities: workers being underpaid; customers being fleeced; environmental or social externalities being ignored; or monopolisation of intellectual property.

Profit should be treated as a sign that the system is in need of correction. Where it arises, it should be redirected to workers, reinvested in better service, or in the last instance, taxed, in order to channel it towards more socially useful and democratically chosen outlets. It shouldn't be the central force driving our economy. It certainly shouldn't be celebrated as an end in itself.

There's a joke that comes from the domain of technical customer support, involving a made-up error code called PEBKAC. Problem Exists Between Keyboard and Chair. A euphemism for "the user is doing it wrong". Anyone who's ever seen someone struggle to use a system they've designed knows the feeling. *The user wasn't supposed to do that!* Or, *It works on my machine*! Your instinct is to defend your design choices; it's the user, not the system, that's at fault.

There is, of course, always an alternative explanation: sometimes the system is wrong. Maybe the assumptions of the designer were wrong, or maybe the situation changed; either way, the system is failing to serve the people it was meant to serve, and that's a problem.

Discussions of our current economic climate tend to evoke a PEBKAC-like response from capitalism's apologists. Any reported problems, when they are even acknowledged as such, are clearly user errors. Millennials drowning in an ocean of student debt? Workers being paid unliveable wages and forced to work under gruelling conditions? People getting priced out of their hometowns? Clearly all these people are just doing capitalism wrong. They should have hustled more, stopped buying lattes, learned to code.

It's an understandable impulse. If the economy is working for you, it's comforting to believe that your success came through your excelling within a reasonable system. Your success is deserved because you worked hard for it, and in fact, you deserve it more precisely because of all the hardships you faced. Those who are unable to overcome those barriers have only themselves to blame, and they too deserve what they get. This is the natural ideology generated by any hierarchical system: those on top are always led to believe that the system is fair and that they're on top because they deserve it. After all, the entire system is devised to ensure their buy-in, and to silence the voices of those who are not on top.

But what if the system isn't fair? What if the problems are not merely due to the personal failings of a few errant individuals, because the system's very design prevents a great number of people from succeeding on its terms? What if the system forces those from the bottom of the economic pyramid to jump through cruel hoops for even the barest whisper of success, making them whittle any trace of excess joy from their lives until they become the ideal efficient worker?

Meanwhile, those on top are showered with luxuries and second chances. For those who meet Silicon Valley's definition of merit, there's no shortage of opportunity to continually fail upward. Your company can lose millions of dollars, you can get fired for creating a toxic work environment, and you can even run your company into the ground, but you're still entitled to the Silicon Valley Basic Income — i.e., more money than anyone would need to spend in a lifetime. Here's your golden parachute of a severance package, and here's a blank cheque for your next venture. No one's coming to repossess your Four Seasons penthouse.

The result is a massive disparity between life chances. Some are coddled and given no end of chances to reach their potential; others are means-tested and subjected to a constant state of despair, never getting the opportunity to realise their potential because society has already decided they aren't worth it. Whatever positive outcomes this system is meant to be producing can only be assessed relative to its corrosive effects on the social fabric.

And yet, even in the midst of a backlash against the tech industry, some of the most powerful people in tech still seem to believe that they're the good guys. Perplexed by the sudden wave of scepticism, they invent numerous reasons to disregard the criticism, which they see as unwarranted and misguided. The critics are just jealous because they're not doers. They're writing clickbait to sell papers.

There's an intoxicating narrative here which seems to fit the arc of some of the very people I used to idolise — the ones whose writing got me hooked on tech in the first place. It's the age-old story of the underdog, and it goes something like this. They start out with nothing, but persist despite the odds, studying and hustling and innovating until they achieve success under the system as they understand it. And it's a *deserved* kind of success, the kind they should not have to apologise for. So what gives people the right to criticise them now? Didn't they do everything right, following the upward path they were supposed to follow?

In their head, they're still the underdog. They're the protagonist of the story. Being the good guy, making the world a better place, deserving their success — these ideas are all crucial to their self-image, and so any attempts to question the fairness or the efficiency of the system feel like a personal attack. They're securing the future of humanity, and they don't understand why people won't just shut up and let them do it. The critics must be motivated by malice or ignorance.

They don't want to consider the possibility that the critics understand exactly what they're doing and just know that it won't be in everyone's interest. They don't want to acknowledge the myopia that comes from their class position, making them an unreliable narrator of the efficacy of the system or the utility of their own role within it. They certainly don't want to admit that their idea of making the world a better place really means *better for them*.[12]

What this means is that those who have managed

to accumulate resources in the current system are not necessarily the best ones to make decisions about where to allocate them. As much as they would like to believe otherwise, their view is tinged with their own self-interest. To borrow a statement that one tech billionaire has used in a different context: when tech luminaries defend their industry, we should take them seriously, but not literally. Their statements are attempts to make their own truth — marketing, in other words. And we should be doubtful when they tell us they're providing value, or liquidity, or efficiency, the way we should be of the propaganda that power always generates in an effort to justify its position.

The peculiar thing about Silicon Valley elites is that they like to think they're different from others. The power and wealth they've accumulated is not enough; what they really want is to be loved. They believe themselves worthy of praise, and they're upset when their critics attempt to hold them accountable for things they don't think they should be responsible for. How typical, they scoff to each other, that the media treats every startup with either scepticism or adoration before ripping it apart as soon as it gets big. How hypocritical.

But a brighter spotlight is a function of growth. The more wealth and power a startup has, the more people come to depend on it; it shouldn't expect to simply abdicate responsibility for customers or workers or communities when convenient. Nor would it be reasonable to expect everyone else to meekly defer to their right to rule. Without criticism, how could the system be held to account? Should the powerful be the sole arbiter of the validity of their power?

After all, there is a simple reason that people tend to root for the underdog. Being an underdog is more than simply a personal identity — it is a relationship to power.

When a startup is young, sometimes it actually does start out as the underdog, competing with an ossified corporation or industry that is in dire need of disruption. Sometimes the startup does contain the seed of a better alternative. The problem is when these underdogs start to grow bigger. They may have begun their journey with a critique of the powerful, but it's an essentially static critique that omits an explanation of *why* these powerful companies got to be the way they are. The establishment — the banks, the hotel chains, the taxi companies — are the bad guys, and the scrappy founders are the good guys, and that's how it will be forever.

There's something tragic about the combination of hubris, ambition, and naivety of those who set out to build alternatives to what they see as the corruption of the old world, only to manifest the same issues in due time. Even as they edge out the incumbents, and even as they become the ones adopting anti-consumer practices[13] and lobbying governments for preferential treatment[14] — the tech disruptors stubbornly maintain their identity as perpetual underdogs, which means anything is justified in their quest to build power. They don't realise that their founding myth of disrupting the powerful no longer applies when they become the powerful ones themselves.

For a long time, I clung to the idea that Silicon Valley was morally better than Wall Street. Even as my tech optimism began to crack and the two industries got more interdependent, I still firmly believed that they were qualitatively different. At least Silicon Valley *tried* to be a meritocracy, unlike the old boys' network of Wall Street. At least Silicon Valley *tried* to do good in the world, unlike those bankers who were motivated by greed. It's kind of funny, then, to discover that people on Wall Street say eerily similar things about their industry that Silicon Valley apologists say about theirs.

Karen Ho's book *Liquidated* — an ethnography of Wall Street, published soon after the onset of the 2008 financial crisis — is a marvellous case study in the myths that people will tell themselves to justify what they do. Some on Wall Street would say that it *is* a meritocracy, where no one cares about your race or gender or family name as long as you're smart and willing to work hard. And some would say that even if people on Wall Street get to make a lot of money — and, yes, are sometimes driven by greed — the industry as a whole is a net positive, because it provides liquidity and discipline to the rest of the economy. Sure, there are occasionally unfortunate consequences, like mass layoffs after a company has been taken over by private equity and asset-stripped, but that's just the cost of efficiency. And anyway, look at all these smart people with prestigious degrees repeating the importance of maximising shareholder value — they can't all be wrong.

In the wake of the financial crisis, disillusionment with Wall Street spread pretty quickly. It became clear that all the rhetoric about Wall Street's positive impact on the world was more marketing than fact, and people stopped believing the story that Wall Street told about itself. We couldn't rely on those same institutions to get us out of that mess — they were too invested in the current system being the way it is.

Now, I think people are starting to feel the same way about Silicon Valley.

The problem with the tech industry — and the reason its rise has been so beneficial to Wall Street, even as individual companies claim to be disrupting the banks — is that it's part of the same system that gave rise to Wall Street. Financial incumbents may occasionally find themselves disrupted, but Silicon Valley is not disrupting the power of finance as a whole: it's not challenging financialisation[15], much less reducing the power that capital holds over our lives. Rather than decentralising power in a meaningful way, Silicon Valley aims to dethrone the powerful in order to take their place.

And this is not a problem that can be fixed simply by having more adults in the room. The problem isn't that the industry is run by unsupervised nerds making bad decisions with impunity; the problem is that any possible supervision only exists within a narrow band of possibility. The problem is the structure in which decision-making power rests with such a small group of people, with such a small window for accountability. Capitalism as usual won't fix the problems, because the internal feedback mechanism is inadequate by design — those who are served poorly by the system are dispossessed of their capacity for political contestation. The only way out is a rupture of some kind, a disruption of the status quo.

One way to view the current stage of our economic system is as adolescence. Capitalism may have been an improvement on its predecessor, but that doesn't mean we need to keep it around forever. No mode of production exists in a vacuum, after all; a good one is designed to suit the needs of the world in which it exists. Sometimes things change, and you have to adjust the rules to fit the material needs of the current conjuncture. Adolescence is a necessary stage, but eventually you have to grow up.

Here's what we have now: a rapidly diminishing natural world, where unfettered resource exploitation is on track to destroy the ecosystems that sustain us; and a broken social order, where some have more resources than they could use in a lifetime while others have almost nothing. And yet, our mode of production tolerates all this because its priority is encouraging corporations to do whatever the hell they want as long as they remain profitable.

Things have gotten to the point where the focus of our economic system seems increasingly misaligned with social need — the positives are getting diminishing returns, and the downsides are piling up. Perhaps it's time to change our priorities and move over to something new.

It's 2019, and I now live in San Francisco. There was a time when I wanted nothing more than to move here as part of the growing wave of tech professionals. Now, though, I have no inclination to work in the tech industry; instead, I'm writing a book calling — only somewhat jokingly — for its abolition.

It feels odd to be here again after my whiplash-inducing personal transformation. I don't see things the same way anymore, and I don't want the things I used to want. I walk past the offices of venture capital firms and where I might have once responded with reverence, I now only feel disgust. The constant startup billboards for cryptocurrency exchanges or marijuana delivery apps are sickening. Parks and transit centres that would otherwise be perfectly lovely are emblazoned with the logos of tech giants, serving as constant reminders of a system that prioritises corporate greed over public good. And most of all, that all this is shot through with flashes of staggering poverty is infuriating.

Whispers of another recession abound. Whatever happens, it will likely hit younger generations hardest; an *Atlantic* article reports that the next recession will destroy millennials.[16] But that pain won't be shared equally — some millennials will fare better than others. The piece quotes a Credit Suisse report on the nature of global wealth distribution, which acknowledges that millennials are less financially well off than their parents overall, but suggests that a tiny proportion of them could overcome their generational disadvantage. Those overcomers are described as "high achievers and those in high-demand sectors such as technology or finance".

When the floodwaters rise, the drawbridge pulls up, leaving everybody outside the castle to drown. When Credit Suisse talks about those working in high-demand sectors, we all know that they don't mean *everyone* working in those sectors — it's a euphemism that everybody accepts. We know who actually benefits from the money in booming sectors; we know that every sector has a divide between the well-paying jobs that will insulate you from economic insecurity, and the poorly-paid jobs that will never get you inside the castle.

One afternoon I share a Lyft Line with a passenger who's on his way to the warehouse of a lesser-known gig economy company, where he works as an independent contractor. He explains that he got called in for his shift last-minute, and he's taking a Lyft because public transit won't get him there in time. As we cross the Bay Bridge, sunlight glinting off the canted suspension cables, the passenger waxes poetic about how much he loves his job. They provide dinner before the night shift! The pay is above minimum wage! Once he got injured on the job and was paid for the entire day even though he got to leave early! He sounds genuinely jazzed about his employer. Our driver looks intrigued and asks for details.

I think about the monstrosity of a society where it's accepted that most companies do the bare minimum for their workers, and it's considered a delightful surprise when they do even slightly more. I think about the gig workers who have been killed on the job, whose families are apparently owed nothing by the company that profits from their work. And I think about the shareholders of these companies — founders, investors, early employees — who are going to cash out massively when their company gets acquired or goes public.

As the tech industry booms, so does the Bay Area's housing crisis. Houses aren't being built fast enough to counter the increase in demand by homebuyers and would-be landlords looking for a safe investment vehicle; in 2019, the median home price in San Francisco was $1.3m for condos and $1.7m for single-family homes.[17] At the same time, the city counts between 8,000 and 10,000 homeless residents, depending on how you define the term; the broader Bay Area counts over 20,000.[18] There are no numbers on how many people who would have preferred to stay here have been priced out.

This is what happens when global capital, funnelled into a local industry, meets a housing policy landscape that is simultaneously pro-developer and anti-development. In a city where the average one-bedroom apartment rents for over $3,000 a month, and new builds tend to be luxury buildings marketed at the upwardly mobile, most units being built are completely out of the reach of anyone without family money or previous startup success. Though developers are required to set aside a certain percentage of new units as affordable housing — usually around 20%, waivable by paying a small fee — what counts as "affordable" in San Francisco is a little unreal.[19] A family of four that wishes to qualify for the program must be making under the area median income of $118,500, and also must have enough saved to afford the down payment on a several hundred thousand dollar home. And even with those restrictions, there aren't nearly enough houses to satisfy demand — in 2017, there were 85,000 applications for only 1,210 units.

Most of the housing debates here bounce between two poles, NIMBYism (Not In My Backyard) and YIMBYism (Yes In My Backyard). The mainstream assumption is that you're either a pro-developer YIMBY, or you're selfish and don't want any new houses to be built. The pro-developer argument goes something like this: if developers are permitted to build enough new units to meet demand, then supply and demand will equalise, and prices will stabilise; eventually, the housing crisis will fix itself. In the meantime, it's sad that people are being priced out of their homes, but the best way to fix it is to allow the market to do what it does best.

But what is the market really good at optimising for? Given the current excess of global capital, the market's most pressing task is finding new investment opportunities. From the point of view of the market, housing is primarily an asset and only secondarily a place to live — exchange value takes priority over use value.[20] Buying a house is considered an uncontroversial investment vehicle: you rent it out to pay the mortgage, then flip it to reap that appreciation when you need liquidity again. The more precarious and price-gouged your renters, the more liquid and lucrative your asset. Capital doesn't shed tears when people are made homeless or impoverished in the process of trying to keep their home.

No wonder the homeless population keeps growing here, just as it is in other cities where housing's status as an asset attracts global wealth looking for a safe place to park.[21] After all, housing can't be simultaneously an asset *and* a human right; safeguarding capital's right to a return requires the ability to evict people if they don't comply with capital's demands for a tithe of their income. Capitalism governs through a mix of carrot and stick, and this particular city — the epicentre of capitalism run amok — is a sordid blend of tantalising dangled carrots and very sharp sticks. There are plenty of opportunities to become fabulously wealthy, but it's also very punishing to be poor.

It's not all bad here, though. Walking around the city reveals defiant enclaves of hope that have resisted displacement despite wave after wave of capital. Murals about solidarity and resistance adorn walls in the Mission district, and Coit Tower still has its New Deal-era murals depicting the 1934 general strike. Wedged between startup offices in SoMa and the Mission are union offices and hiring halls, a memento of a different chapter of labour history. Schools and streets bear the names of labour organisers and social activists fighting for a different world than the one in which they grew up. And signs of support for workers who have unionised at a local brewing company are sprouting in the windows of seemingly every bar in the vicinity.[22]

As tiny as these seeds are compared to tech's leviathans with their fountains of money, it's a nice reminder that things don't have to be this way. If you're disillusioned, you're not alone. Silicon Valley's tech-inflected brand of capitalism is only the latest battleground in a long war; others are doing what they can to push back, and they've been doing this for a long time.

May of 2019, there's a protest outside Uber's headquarters in San Francisco. The immediate occasion is Uber's upcoming IPO, which is projected to bring the company to a valuation of nearly $100 billion. If it reaches that, Travis Kalanick — a co-founder and former CEO who departed the company after a wave of scandals in 2017 — will mint several billion dollars. Garrett Camp, the company's lesser-known co-founder who hasn't worked for the company in years, will also be a billionaire. Early investors will make back several times their initial investments; the executives will make many millions; lower-level employees might become millionaires, depending on how long ago they joined and how valuable the company finds their work.

According to their S-1 filing, the company has over 20,000 full-time employees and almost four million active drivers. Of the money raised from the IPO, only 3% is set aside for drivers — significantly less than the amount reserved for the company's erstwhile CEO. And yet these drivers have been dedicating their waking hours to creating this company's wealth, at wages they cannot negotiate, and which mostly seem to be going down; they don't have employer-provided healthcare, expense accounts, catered lunch, or even bathroom breaks. And only about a quarter of the active driver base will be eligible for any reward from the IPO at all.

This protest is nominally about Uber, but it's clearly about more than just that. There's a pent-up rage here, about workers' rights, about the tech industry's fuelling of inequality, about the cost of living on this absurdly expensive peninsula. Some of the protest signs mention current Uber CEO Dara Khosrowshahi: his $17m mansion in Pacific Heights, his yearly salary of $43m in 2018, the millions he's expected to make during the IPO. Contrary to the tech industry's typical celebratory attitude towards massive riches, the message conveyed through the signs and chants is that the lopsided wealth distribution is not legitimate — that the ludicrously large payouts to shareholders come at the expense of the drivers living out of their cars in order to stay afloat.

There are a few Uber and Lyft drivers at the protest, but more are on the streets, working; many of the cars with Uber or Lyft stickers honk in support as they drive by. The contingent of actual drivers at this protest seems dwarfed by their supporters: a radical marching band which often shows up at events like this, a variety of activists and organisers, progressive politicians. There are even a few white-collar tech workers from other companies, some of whom I've previously met at other events. It doesn't look like there are any Uber engineers here, but that's understandable; publicly protesting your employer is not a good look in this industry. The only public show of support for this protest that I've seen from Uber employees is an anonymous *Medium* post published a few weeks prior.[23]

The stifling culture of secrecy and familial loyalty within the tech industry masks deeper problems of worker solidarity. Uber employees may hesitate to speak up in support of drivers out of a fear of reprisal, but it's unclear how many of them support drivers' struggles at all. Why, after all, would they identify with the people protesting their company over the company that cuts their paycheck? And in a larger political climate where workers who ask for more are seen as entitled, and "job creators" are seen as infinitely more deserving than the workers doing those jobs, why would they have any sympathy with the small number of malcontented drivers fighting for something as irrelevant as "workers' rights"?

And yet, Uber employees may have more in common with drivers than they know. A few months after the protest, Uber announces a round of layoffs comprising approximately four hundred employees from its global marketing team[24]; this is followed in September by a further four hundred employees, this time from its product and engineering teams.[25] These white-collar workers may be in better financial shape than the average rideshare driver, and most of them will probably find new jobs without much difficulty, but losing your job is losing your job, and it can be especially tough if you rely on your employer for health insurance or visa sponsorship. The fact that companies like Uber can lay off full-time employees with impunity is not unrelated to Uber's ability to raise so much money for an app whose primary innovation is a new way to exploit workers. In an atmosphere where capital is powerful and labour is weak, all workers are vulnerable, even if some are currently more vulnerable than others.

Something you notice when you walk around San Francisco is that the city has two main bikeshare systems. One system consists of blue bikes docked on the side of the street; the other system is dockless, consisting of orange bikes haphazardly abandoned on the sidewalk. If you were unfamiliar with the city's embrace of unbridled privatisation, you might assume that both bikeshare systems were owned by a public transit authority, with the docked bikes being the main service, and the dockless bikes there to fill the gaps. But this is 2019, and we're in the epicentre of technocapitalism; opportunities to provide useful services for the public good are quickly snatched up by tech companies in the pursuit of profit.

The orange bikes are owned by Jump, a scooter and electric bikeshare company founded a decade ago in New York City and recently acquired by Uber for an undisclosed sum. The blue bikes are owned by a company called Motivate, which also manages bikeshares in several other major US cities, and which was recently acquired by Lyft for a rumoured $250m.[26] Reportedly, Uber had been eyeing the company as well — both companies are trying to dominate the transportation space — but passed in favour of Jump, due to concerns over the cities' power to cancel contracts and the fact that some of their workers are unionised.[27]

For tech companies, acquisitions are rarely a matter of buying successful businesses based on their price-to-earnings ratio; it's about synergy, company strategy, the road to domination. Motivate may or may not have been profitable, but the reason a ten-year-old company with minimal assets and non-tech profit margins was able to command such a high valuation is because a different company, which also does not have software margins but has convinced investors it is tech-adjacent, saw Motivate as part of their path of expansion. The latter company was deeply unprofitable, but it had the faith of various investors who believed that Lyft would eventually command enough market power to profit through a combination of gouging customers and underpaying workers.

The upshot of this bizarre world-spanning economic saga is that Lyft now owns the bikes I occasionally ride around San Francisco.

San Francisco isn't the safest city for biking, but other public transit is limited or unreliable, so I spend a lot of time getting around via the blue bikes. Once in a while I encounter issues with docking — sometimes the docks nearest me are empty, so I have to walk; or, I'm trying to dock my bike but the spaces are full, so I have to take an annoying detour. Motivate employees occasionally move bikes around to minimise passenger inconvenience, but it's never enough, and I don't know if it's because they're understaffed or because they don't have the right data.

I'm weirdly fascinated by the technical problem of routing bikes around, and I brainstorm how I'd solve this if it were my job. You could track variables like the weather, the day of the week, the time of day, the location of nearby events. You could anonymise and aggregate existing customers' typical travel patterns and feed that into the algorithm too. It's a minor thing in the grand scheme of things, but some days when I'm circling a five-block radius looking for a place to dock my damn bike, I fantasise about working on a problem like this. It would be nice to work on something that solves a real need — a fun technical problem that could actually improve people's lives. A little piece of uncomplicated good in this messy world.

But given the current state of things, that would feel like refuge. I can't do that again, not now; all my profiling leads me to believe that we need systemic change. I can work on docked bike rotation algorithms after the revolution.

What to do in the meantime? The tricky thing about trying to challenge the current system is that it's stunningly good at containing potential opposition. There's a reason it persists despite all the misery it generates: it's structured such that most attempts to change the power balance are easily countered.

What this means is that individual actions to confront the dominance of corporations are limited in their power. You can boycott Facebook's main product, but that's not going to stop them from spending billions to scoop up other social platforms and then merging their infrastructure together in anticipation of impending antitrust action. And it's also not going to dislodge their key role in the digital advertising ecosystem — they've already leveraged their way into a powerful position, one that extends far beyond the reach of their original product.

Even in the workplace, individual actions have limited impact. You personally refuse to work for a company that collaborates with ICE? They'll still be able to recruit employees who either don't understand the moral considerations or who don't care, and the products will still get built.

That doesn't mean that these actions aren't worth doing, of course. Every movement has to start somewhere. But if your goal is actual change, then these actions can't remain on the individual level — they have to be collective. Personal lifestyle changes can be important, but there also needs to be a broader movement that represents a substantial challenge to the system. And the way you get there is through organising — getting other people on board with your mission, in order to extend the power of your actions beyond the scope of your personal leverage.

None of this will be easy. The overarching structures determine what acts of resistance are feasible, and some choices are already all but made. Boycotts can be highly inconvenient for customers; after all, some companies have flourished precisely because it would be hard *not* to use them. And withholding labour — through strikes, sabotage, or quitting — can be more likely to hurt the worker than the company. Whether it's worth doing depends on one's leverage, risk, and level of coordination with others.

The upshot is that there are no solutions in piecemeal. Even the things that look like promising islands within the capitalist ocean — like the free software movement or platform cooperatives — exist on a tiny scale compared to the privatised tech industry. In our current world, we don't have a comparable alternative; full-fledged alternatives currently only exist in the imagination, and each would have its own downsides. But maybe it's time to take a leap of faith anyway: out of the flames of our legacy codebase, toward the possibilities of the unknown.

ELEVEN: A NEW INDUSTRIAL MODEL

We need to be ready for a world with trillionaires in it. And that's always going to feel deeply unfair. It feels unfair to me. But to drive society forward, you've got to let that happen.
— Sam Altman, President of Y Combinator, in a quote for *Business Insider* in 2016[1]

So now onto the central question animating this book. What does it mean to abolish Silicon Valley? What would a better system for developing technology look like?

My stance starts from the axiom that we need more than mere geographic diversity, where the Silicon Valley model is duplicated in Silicon Alley and Silicon Beach and Silicon Wharf. And we also need more than a few personnel changes on top. More geographic and demographic diversity might be a start, but that's not nearly enough. What we need to do is to reclaim our world from capital.

I don't have a numbered list of concrete demands. All I have are inchoate sketches of how the world could look, drawn out of the horrors of the world we have today. Some will seem impossibly radical, but the point of making a demand is to put a stake in the ground — to anchor the imagination. Being able to envision utopian alternatives may pave the way for more moderate possibilities that would otherwise seem unachievable.

None of this can happen overnight; some of this can't happen in a decade. This is a long project that will require a radical restructuring of existing institutions and of the way we relate to the world as individuals. The path toward building a better world is always only partially complete; there will always be further to go, a horizon that can't even yet be imagined.

Abolishing Silicon Valley means freeing the development of technology from a system that will always relegate it to a subordinate role: that of entrenching existing power relations. It means designing a new system that isn't deluged in the logic of the market. It means liberating our world from the illegitimate reign of capital.

Perhaps this sounds unfair to capital; perhaps I sound like I'm not grateful enough for everything that capital has given us. But we don't owe capital anything. The things we attribute to capital were built by workers: people who laboured and sometimes died in the process, their contributions unrecognised in death as in life. So don't thank capital — it doesn't deserve our gratitude, and it doesn't need it, anyway. Thank the people who created everything that capital always takes credit for. Capital is a means of accounting for wealth ownership, not its creation, and that means it's perpetually shrouded in a fundamental untruth. Time to leave the swamp of capital behind and start over with something new.

1. Reclaiming Entrepreneurship

One of the defining features of Silicon Valley is growth entrepreneurship. Software is usually involved to some degree, mostly because software is helpful for scaling quickly, though many companies we associate with Silicon Valley nowadays go far beyond selling software. The process typically goes something like this: some people have an idea, secure funding, build it, then scale it up. There's lots of variation in the details, but there's always a common thread of raising money that is expected to make a return. The return could be achieved through getting acquired by a bigger company, ideally for more money than the most recent valuation; it could mean going public. Either way, the big money is in companies with the potential for massive returns.

Who you take money from determines who you serve. Investor money is a gift with strings attached. You may have some leeway in how you run your business, but at the end of the day you're still accountable to your investors. They might ask you to acquihire a failing portfolio company or make strategic product decisions based on what's best for them, and even if you think their advice is bad for your company or for society, their needs will prevail. After all, it's *their money*. Who could argue with that?

So let's imagine a new model of entrepreneurship: entrepreneurship for non-capitalist ends. Instead of entrepreneurship as primarily a private endeavour, accountable to private shareholders, we could have entrepreneurship as a public service. It would be driven by the aim of creating something useful, rather than the prospect of unimaginable wealth, and the rewards would accrue to workers and the public in a much fairer split than we see now. Founders and executives and investors should not become billionaires — or, God forbid, trillionaires — while their workers get next to nothing. The reward for building a successful startup should be the privilege of building a successful startup, within a larger economic environment where massive wealth is not needed to secure financial stability.

One piece of this is to create a publicly-owned investment fund whose scope is limited to non-profit ventures. This would be similar to what's traditionally considered "social entrepreneurship", with socially beneficial aims and no pressure to turn a profit, but in this case there would be ample access to funding. The government could provide the upfront capital and take majority ownership but otherwise get out of the way of daily operations — similar to how most investors work now. Pitch decks and financials would be judged on the basis of practicality and social value rather than market domination or profitability, and fraud or malfeasance would be severely held to account to dissuade Theranos-style grifts. This fund could also provide the necessary capital for existing corporations whose workers would like to convert it into a worker-owned co-op, or another suitable structure enabling worker control of production, as in the Lucas Plan of 1976[2]. Such an investment arm would be a way for the government to take a more active role in steering industrial strategy, rather than leaving it all to the whims of private markets.

Doing this right would require more than simply having the government fund a few non-profits. The current startup landscape is defined by the needs of investors seeking a return on their capital, and that affects the culture downstream. After all, few would-be entrepreneurs would consider this public service variant if the private alternatives were laxer and more financially rewarding. To fix this, the broader industry needs to be transformed beyond recognition, so that it's no longer possible to become fabulously wealthy through entrepreneurship. The goal is to change the culture so that technology entrepreneurship becomes an endeavour guided by the idea of public service, where technology is scaled up in a non-profit-driven environment free from concerns of moats or domination. Profit-driven efforts to develop technology should be seen as suboptimal, and they should be converted into public variants when useful and feasible.

Ultimately, this would require eradicating the roles currently played by the institutions that invest and govern on behalf of shareholders. Existing investment firms should be seen as following legacy specifications, and they should be either repurposed or replaced by new institutions with less profit-driven charters. Reclaiming entrepreneurship from capital does mean eliminating private ownership, at least when it comes to larger enterprises — a company that's able to mint billionaires is too large to be run for the benefit of a small number of shareholders. Concentrated private ownership of corporations is deeply entwined with our landscape of heavy wealth inequality, so addressing the latter will require new modes of corporate ownership: more non-profits, or ownership distributed among relevant public agencies and workers, depending on the circumstances. Companies that need to raise more money to expand could issue bonds or turn to public funding rather than equity markets.

Now, government services tend to have a bad reputation when it comes to technology, especially in the US. But that seems to me like something of an attribution error — there's no end of examples of private companies being inefficient or careless; we just don't attribute their flaws to private enterprise as a whole. The reason there have been more visible successes in the private sector has to do with the larger political climate in which private enterprise is given more funding and room to manoeuvre, whereas many governments have turned to cost-cutting and outsourcing measures as part and parcel of neoliberal policy. In any case, government funding does not necessarily mean the government as it is now, as it exists in a capitalist system; we should aim for a more democratic form of government, actually accountable to the people it is ostensibly serving, and this new form of entrepreneurship would be downstream of that.

Of course, public funding isn't enough to guarantee success. If a non-profit has to compete with a better-funded, for-profit variant, the latter is more likely to win as long as the field is skewed in favour of capital: it could subsidise costs to win over customers; it could underpay workers to save on labour costs; it could hoard code or data; it could dump money into marketing to ensure brand recognition. Other efforts to level the playing field will be needed, ranging from tightening labour protections, to mandating some form of public disclosure of data and IP, to ensuring that the public fund takes a stake in all relevant companies, all the way to outright banning for-profit competitors.

This new form of entrepreneurship probably sounds quite strange, not least because it goes against the most prominent manifestations of entrepreneurship that we see today — i.e., the unicorns of Silicon Valley. But that's intentional. Our current understanding of entrepreneurship is deeply saturated in power-hungry capitalist greed, leading to undemocratic control of the technological infrastructure that underpins our lives — not to mention massively wasteful economic inequality. Whatever merits the existing system of private entrepreneurship may have had, we're now brushing up against its limits, and it's time to consider something new.

2. Reclaiming Work

Beyond the initial act of entrepreneurship, we should rethink the next stage: work. Right now, capital has inordinate control over many aspects of work, including the conditions of work, how it's compensated, and how its terms can be contested.

In the tech industry, there has emerged a broad division in the workforce. On one side, there are the highly-valued workers building the product and making the decisions; on the other side is everyone else. This division has emerged due to capital's tendency to keep the means of production close by, while outsourcing other jobs as much as the market can bear. To fix this, a number of avenues need to be addressed.

The first goes beyond just the tech industry: workers need more power. They need a higher minimum wage and raises that at least keep up with inflation. They need universal healthcare and unemployment benefits so they can quit a job that is exploitative, toxic, or immoral. They need unions that protect them as workers so they're able to blow the whistle in cases of corporate malfeasance and organise around better working conditions. Migrant workers need more lenient immigration regimes so that they're able to organise or switch jobs without being entirely dependent on the support of their employer. And as long as corporate boards exist, workers need seats on them, ideally several, through democratically elected representatives.

The second avenue has to do with income inequality. This issue is especially pressing in the tech industry, where the difference between a CEO's income and that of the lowest-paid worker on payroll can be enormous — among large US corporations, this ratio is typically several hundred.[3] One way to address this is by pegging the CEO's total compensation to that of their lowest-paid workers, and instituting a maximum wage based on the lowest wage, including workers who are classified as independent contractors rather than employees. This maximum/minimum pay ratio could start high to reflect existing reality and shrink as quickly as the landscape will allow.

One particular quirk of the tech industry is that workers in non-management roles often have access to unexpectedly high compensation, most of it via stock grants. But stock is typically highly unequally distributed among employees, and after a certain point this becomes a clear economic inefficiency. This could be addressed through lowering the maximum wage to a more reasonable amount while ensuring a more equal distribution of stock, or, for publicly-funded companies, removing stock grants for employees entirely. And for technical roles, public salary transparency based on posted bands would be helpful in order to minimise discrimination, ideally coordinated through industry-wide and location-specific bargaining.

However, minimising income variation within a workplace can backfire if it merely directs money toward shareholders rather than workers. These efforts would have to be in tandem with other approaches to prevent capital from reaping the rewards. Companies with excess profits accrued through rent-seeking should be considered targets for nationalisation. Failing that, they should have to lower prices, increase R&D spending, and/or raise wages; the rest could be taxed away. And a strict wealth tax would also be useful given that those with a high amount of savings can draw passive income solely from investing. The overall effect would be a fairer distribution of wealth: higher living standards for those currently on the bottom, accomplished through curbing the excessive power of the ultra-rich.

The third avenue has to do with the contractor workforces of tech companies, some of which outnumber their corresponding full time workforces. The ones who are working full-time should be given a clear path for converting to legal employment; those who choose not to convert should be paid a higher net hourly wage than employees, given that they're not provided benefits. And when it comes to gig workers who don't want to be full-time employees, they should have greater control over what jobs to take — no threats of deactivation if they skip too many jobs, nor jobs specifically chosen to get them working longer hours. They should have greater ability to negotiate pay, individually and collectively, and there should be a more transparent, worker-friendly process for resolving disputes with the company.

The fourth avenue involves control over technology used in the workplace. The gig economy may be the purest expression of software as a means of extending capital's power over labour, but this phenomenon is present in other industries too. Fast food, hotels, factories — any push towards automation under capitalism is driven by the desire to save money, even if it means stressing out employees and shifting work onto dissatisfied customers. To go the other way, workers need more control over how technology is developed. Technology should never simply be foisted on them to help their employer extract more value from their labour; the technology should grant workers agency, and its aim should be making workers' lives better. Workers need substantial say over the development of any new technology to be introduced in the workplace, and they should be actively involved in its design, in order to ensure that it fits their own needs. Technology that only serves the needs of management, rather than workers, should not be deployed at all.

The final avenue entails changing the way software development is done. As technology gains an increasingly crucial role in society, we need rigorously vetted software that can withstand audits, written by people who are qualified to write it. Software engineering could be licensed, analogous to how doctors or lawyers are licensed, starting with roles that could present the greatest downsides if done poorly. This would help enforce minimum code quality standards, which is especially important when it comes to user privacy and security. This licensing process could also come with the software version of the Hippocratic oath, which could serve as an ethical baseline for when to blow the whistle. This could be enacted through governments and other organisations refusing to procure software from companies without certified engineers in certain roles, or legislation forbidding companies above a certain headcount or revenue from hiring uncertified engineers. And to ensure that this isn't a regressive policy that locks out would-be engineers, the training should be available for free and should come with a stipend at higher levels.

Longer term, an interesting vision is the hiring hall model, such as the one that the International Longshore and Warehouse Union organised for in the past.[4] This would necessitate a different way of developing technology, with knowledge going back to the community rather than being trapped within a company. People would work for companies on a project-by-project basis, then would go back to the hiring hall for the next project. One beneficial cultural effect would be to reaffirm that workers' loyalty should be to their class, rather than to their employer.

Ultimately, the aim of reclaiming work from capital is to create a radically different vision of work. Rather than the current system that governs through excess financial incentives for a select few and coercion for everybody else, work could be reformulated as a more democratic endeavour whereby we collectively create the things we want in society. This could mean more worker co-ops, stronger unions, and greater public input into industrial strategy, in line with the UK Labour Party's proposals for democratic public ownership[5].

3. Reclaiming Public Services

One major factor underlying the success of many tech companies is the background phenomenon of privatisation in many parts of the world. Accordingly, a whole landscape of sectors which *could* have been public services are instead easy prey for tech companies: healthcare, education, banking, mobility, community, and housing, to name a few. To reverse this, we need to excise capital and restore public ownership. The Universal Basic Services model as proposed by University College London's Institute of Global Prosperity in 2017 could be a solid framework, though we should go beyond the breadth and extent of the suggestions in the report.[6]

That means universal healthcare. It means better funded primary and secondary education and tuition-free post-secondary education, with courses geared towards human flourishing rather than merely meeting the skills needed by existing corporations. It means public banking, along with higher wages and lower costs of basic goods to reduce the need for exploitative financial instruments like payday loans.

Mobility in particular needs to be reclaimed from VC-backed startups. Take scooters, for instance: instead of five different tech companies littering them indiscriminately with minimal public oversight, imagine a system that's municipally owned and free at the point of use.[7] Similarly, rideshare apps could be part of local transit systems — publicly funded, positioned as an infrequent alternative to public transit, with well-compensated and unionised drivers.

When it comes to community, instead of WeWorks everywhere, imagine better-funded public spaces like libraries or community centres, free from the strictures of commodity logic. And we should have more democratic control over the online platforms we spend so much time on, through user ownership, state ownership, stronger regulation, or decentralisation. The British Digital Cooperative model, proposed by Dan Hind for the Common Wealth think tank, offers a useful alternative to our current profit-driven landscape of digital media, and would create a healthier and more democratic ecosystem through public ownership of media platforms.[8]

Finally, housing should be seen as a human right, and thus provided as a public good. Social housing should become the norm, and local governments should progressively reclaim housing stock from private hands while also building more if needed. At the same time, private landlords should have their power curtailed, with strict rent controls (including rent caps) so that no one needs to rent out their primary residence on Airbnb simply to afford rent, and so that owners of multiple units are not able to reap excess rents from doing the same.[9]

Overall, the purpose of public funding should be to decouple the funding from the service. Silicon Valley's current focus is distributing resources within a regime of scarcity, and whatever success it has achieved in this area has come at the cost of a highly inefficient concentration of wealth. We should instead manage resources with the aim of meeting human need, creating or repurposing the structures needed to achieve this goal. It shouldn't be necessary to charge for basic goods or services at the point of use — anything that's socially valuable could be funded through progressive taxation. Plus, increasing the number of free public services is a way to tackle capital on the cultural terrain, because it challenges the core ideological tenet of capitalism: the idea that you must pay for everything you use. The end result of expanding public services is to diminish the importance of money by reducing the number of outlets for it, thus improving quality of life for those with less.

4. Reclaiming Intellectual Property

Capital's hold over intellectual property is key to how the tech industry evolved into what it is today. Reversing this requires undoing intellectual property laws that favour corporations: reducing copyright and patent term lengths, tightening the scope of what can be considered patentable, allowing more fair use for non-profit purposes. Patents should be granted to private enterprises rarely — only when there is no suitable public or non-profit entity that could use the patent instead — and should not be privately tradeable for the purpose of patent trolling.

Trade secrets are another form of intellectual property, and these should be reduced in prevalence where relevant to the public interest. Consider the case of self-driving car technology, the subject of a recent trade secret dispute between Google and Uber. Google might argue that its self-driving car technology is their private property, and so for a former employee to share that technology with a competitor would be theft.[10] And yet, if the technology is truly as useful to society as Google claims, then it would better serve the public interest if that technology were owned and developed by an institution accountable to the public. In a similar vein, utility companies have been subject to public takeovers in the past on the basis that their product should be considered part of the commons, regardless of how much it cost them to develop the technology.[11] Public interest should triumph over concerns of private investment.

Google's search index is another example of a trade secret that should be considered public property, as it is part of a societal commons.[12] After all, Google did not create the content in the index; it merely organises the content according to various mathematical calculations. And sure, the end formula may be complex, and it may have taken a lot of engineering effort to get there, but that doesn't mean Google should be able to lock it up and milk it for all it's worth — especially since some of the early engineering work was funded through public grants.[13] The index, and all the algorithms that produce it, should be treated as a public good; Google could be licensed to run it for the public benefit, but it should have to make its terms public, and the terms should be contestable.

Eventually, all private corporations using proprietary technology should have to release it under an open license, though there could be a delay between deployment and releasing the source code, and the priority should be corporations with higher revenue or usage. Beyond weakening corporate power, this would also have the effect of encouraging innovation; coupled with efforts to increase computer literacy, this would give end users more control over the technology they use. A softer version of this, which could be useful as a transitional step, could be to mandate an open API for companies in crowded spaces, so that their intellectual property essentially becomes commoditised — something that is already happening with legislation targeting scooter apps in Washington, DC.[14]

Companies that choose not to make their code publicly viewable and contestable should be held liable for any harm caused, even indirectly. In the US, an amendment to Section 230 of the 1996 Communications Decency Act could make platforms featuring user-generated content accountable for any content explicitly promoted in a recommendation engine.[15] This would be most relevant for YouTube recommendations, which have come under fire for being gateways for extremist content.[16] The equivalent of the Freedom of Information Act (FOIA) for private companies above a certain size would also be useful, so that internal decisions about product or corporate strategy are archived and disclosed upon request.

Data portability is another important avenue for tackling corporations' hold over intellectual property. Data that is generated or created by users is, under the current regime, treated as if it belongs to the company that has captured it, and this reality is reflected in company valuations where user data is valued as an asset. Instead, we should see data as belonging to the people who created it and to the public at large. One example of actions taken in this direction is the city of Barcelona under CTO Francesca Bria, which has modified its procurement standards to enforce public ownership of data within private contracts.[17] Data currently captured by ridesharing or scooter or housing-related startups should be anonymously aggregated for use by urban planners, and any personal data should be easily portable for export to another platform. The latter would also have the effect of increasing worker mobility between gig platforms.

Overall, instead of technology being privately owned and developed in secret by multiple competitors, we should aim to have more open protocols and decentralised services. Secret development makes sense for small entities working on technology not yet shown to be in the public interest, but once a form of technology is mature enough, there should be a public standard, with democratic oversight and open collaboration rather than secretive competition. Technology should be developed primarily outside the confines of for-profit owned by shareholders; instead, it should be developed within worker co-ops, government agencies like the UK's Government Digital Service, publicly-funded research labs, non-profits, and grant-supported open source projects. The resulting products should be released under an open license where possible.

For products and services where meaningful competition is important, the lever of intellectual property reform offers an alternative to antitrust efforts to break up tech companies, which would have limited success in restoring competition due to intellectual property hoarding and high start-up costs in the sector. Rather than breaking up companies horizontally by undoing acquisitions — as has been proposed by US Senator Elizabeth Warren[18] — their power could be challenged vertically, through removing their control over intellectual property. This would essentially tackle their power from the bottom up. Once a product gets enough traction, it should be unbundled into a series of open protocols and standards to permit alternative implementations, analogous to the way web browsers are developed today (though without the corresponding corporate dominance[19]). There should also be high portability of personal data, as well as publicly-released code and other data so that competitors can be built without duplication of work.

5. Reclaiming Culture

Finally, the most nebulous thing to reclaim from capital: culture.

Advertising is a key component of capital's control over culture. When corporations add their names to sports arenas, or embed their commercials as YouTube pre-roll, or pay for product placement in television shows, it's an attempt to ensure that capital is inseparable from any sort of social or cultural space.[20] You're not supposed to enjoy anything without being reminded that these corporations exist and would like you to buy their products.

Whatever net benefits the advertising model may have had in the past, we're now in a situation where commodity logic has gone off the rails, creating new imperatives orthogonal to societal well-being. The ubiquity of vectors for advertising and the temptation to create products that do not address social need have combined to create a world where demand for unnecessary goods is inflated through immense advertising spend — money that could be spent on more meaningful cultural production than an ad for Diet Coke. An economic system that enshrines the primacy of selling commodities can only lead to overflowing landfills and resources being taken from the earth faster than they can be replenished.

We should see advertising as market distortion and find ways to limit it accordingly. Advertising is already regulated to some degree: there are government agencies to police falsehoods in marketing, and there are restrictions for advertising particular goods, like tobacco or gambling. That regulation could be expanded to limit advertising expenditure for goods produced using exploitative labour practices or unsustainable environmental impact. This doesn't necessitate an end to information about commodities — the gap could be filled with better-funded consumer research and independent journalism, as an alternative to the unverifiable and deliberately manipulativ eclaims of advertising. Just as democratic ideals are subverted if the wealthy are able to buy votes in the political arena, so too are the purported aims of the market subverted if corporations are able to effectively buy customers. As long as the market is an important avenue for expressing consumer wishes, consumers must get the information they need to make informed decisions *without* the undue influence of well-funded marketing departments.

On the supply side, steps could be taken to dramatically reduce the outlets available to ads run by for-profit enterprises. This could be done by banning the most intrusive or duplicitous forms of advertising outright — pre-roll ads, full-screen video ads on mobile, native advertising — or at least making it a user-configurable option directly within the product, as opposed to the arms race of_external ad-blocking software we have now.[21] Targeted advertising using personal data or browsing history should be greatly restricted in precision, and when it comes to political messaging, should be a matter of public record. At the same time, ad-supported media platforms that would struggle financially with less ad revenue should be supported through democratic public funding, along the lines of recent proposals by the UK Labour Party in conjunction with the Media Reform Coalition.[22]

Beyond advertising, many problems of capital's control

of culture could be fixed through a combination of loosening intellectual property laws, stronger workers' rights, and more public services. The games industry has been under fire lately for exploitative practices, both towards workers and towards customers; to address this, workers need more control over what gets built and how they build it, which could be achieved through strong unions[23] as well as worker co-ops.[24] Shorter copyright terms would prevent creative monopolisation over cultural treasures — something we've seen, for example, with Disney rising to multi-billion-dollar heights thanks in part to its strategic lobbying to extend copyright terms. And finally, rather than a teeming landscape of companies offering access to movies, television and music for a monthly fee — and taking a massive cut of the proceeds — we could democratise both production and consumption, with better public funding to support artistic ventures, and vastly expanded libraries (physical and electronic) to decommodify access.

TWELVE: EPILOGUE

Growing up, I slowly had this process of realising that all the things around me that people had told me were just the natural way things were, the way things always would be — they weren't natural at all. They were things that could be changed, and they were things that, more importantly, were wrong and should change, and once I realised that, there was really no going back.

— Aaron Swartz, a computer programmer and political activist who was prosecuted by the US government in 2011 for systematically downloading academic journals in protest of intellectual property restrictions; in 2013, he took his own life.[1]

So what now? None of what I've suggested in the previous chapter will be easy, and the overall arc will require coordination from many actors pushing in the same direction. And it's unlikely to be achievable all at once, in a revolutionary act of abolition; utopia can only be built starting from where we are now. The hard part is figuring out how to get there without triggering a backlash, by working with the world as it exists. Progress may have to be slow, and whatever comes next won't be perfect, but neither is what we have now. Stepping backwards can be a good thing if you've been moving in the wrong direction.

The aim of abolishing Silicon Valley is to reclaim our world from capital, which means diminishing the power that money holds over our lives. Getting there would require decommodifying essential goods while also radically transforming the way we think about work — so that people work not because they need to pay the bills, but because they want to do something useful for other people.

Of course, whether this aim is valuable is a highly contested matter. Any attempt to challenge capital should expect resistance. Diminishing the power of capital would mean diminishing the power of those who currently serve as capital's agents, and they've gotten used to the fruits of that power; they will use all the formidable tactics at their disposal to maintain their illegitimate grip. Change is won through struggle, not ideas alone, and it won't be easy. On the other hand, doing nothing isn't really an alternative when that means languishing in a rotting status quo.

In a sense, abolishing Silicon Valley isn't really possible — at least not anytime soon. And yet the point of making the demand is to illustrate that the systems that govern our world are *constructed* — the product of choices by human beings who came before us. Things weren't always like this, and they don't always have to be like this either.

Within the tech industry, the trends seem to be moving the right way, and that gives me hope. Those entering the tech industry now are joining at a time when tech has already lost some of its lustre, which makes it harder to unreservedly drink the Kool-Aid. They've come of age in a world where the tech industry is no longer the underdog, and that changes the tenor of the conversation. The glamour has worn off; the latest cohorts seem more sceptical of the industry's claims to the greater good, more concerned with social impact.

As the industry grows, and the composition of its workforce shifts to include those who were previously excluded, it's headed for a reckoning. The industry's vast wealth and power — always dubious to those looking on from without — are being increasingly called into question by those within. Power without responsibility is not sustainable, and there will be a push to get the industry to either use that power wisely or to relinquish it. Bribing tech workers with lavish perks only works up to a point; for those who want more than the money, less traditional avenues will look appealing, like social entrepreneurship, non-profits, and public policy.

Many, however, will choose to work for companies they won't always agree with, and from this cluster I draw the most hope. They will go in with their eyes wide open, less inclined to take their company's messaging at face value, more likely to treat it as just a job. The money won't be enough to buy their loyalty — they know where the money comes from; they know the human costs of it. As a result, they'll treat the conditions of the workplace — the rules, the conventions, the distribution of power — not as static and immovable, but as contingent, and open to transformation. They'll organise with their colleagues to fight for better conditions and workplace control, counteracting management's power from above with united power from below. They'll support progressive politicians and groups that advocate for workers and tenants against the bulwark of capital. And they'll learn from other industries where workers have successfully organised despite the countervailing forces, and take action in solidarity with other workers.

In a way, it's not fair that the onus should be on workers to take the risks necessary to enact change — and especially not the industry's newcomers, who may have the most to lose. It shouldn't be their responsibility to ensure that their company behaves ethically. And yet, we can't simply leave this crucial task to the people who currently have the power to change things, no matter how lofty their public statements about ensuring the future of humanity. Those who tolerate cruelty toward the people who are under their power now — the workers whose unceasing labour is responsible for all their power in the first place — have already shown that they do not have other people's best interests at heart. They are certainly not worthy of *more* power.

And so the responsibility lies on those who have little power individually but so much power together. What is the alternative when those with the power to make things better consider that to be outside the scope of their job description? The present configuration of power is what got us into this mess in the first place, and the path forward lies in disrupting that balance of power, in order to distribute it anew.

Even now, having written this book, it's still hard to piece together the steps that got me from where I was to where I am now. Maybe that's because there were many moving parts in my journey — it wasn't just one epiphany or trigger, but rather a series of them, prompted by observing the world around me and realising that my worldview no longer held up to scrutiny, first gradually and then all at once. It was like I was driving a car that had been running low on gas, and then all of a sudden it shuddered to a halt.

I used to believe that the world was basically fair, and so I had no responsibility other than to maximise my individual success within it. But somewhere down the line, I started to see the entire system as flawed, unsustainable, unacceptable. I had to unlearn so much of what I had previously believed, because none of it made sense anymore. Focusing on individual achievement within the system no longer seemed reasonable when it felt like the legitimacy of that system was crumbling.

I started this journey wanting to understand the problems I was seeing within the tech industry. In the end, I discovered that whatever horrors I noticed within the industry were merely products of the deeper contradictions within capitalism. Capitalism's fundamental untenability predates the birth of the semiconductor, and its underlying problems extend far beyond the reach of Silicon Valley; I just hadn't noticed them before, because they didn't affect *me*.

And that's not an easy thing to come to terms with. It's not pleasant to acknowledge the injustice of the system you took for granted, and which in any case is the only system you really know. I'm still a product of this system; my deep-seated beliefs, my skill set, and my understanding of personal success are shaped by the world I grew up in. Even as I'm now beginning to piece together a new mental framework, it's a struggle, not a simple replacement. There will always be a part of me that defaults to the bad habits I learned in the current system. But I don't believe in the current system anymore, and so I choose to listen to the part of me that itches for something better.

Still, I'm not exactly sure what that looks like, and I can't really fault anyone for choosing to stick with the familiar. This book was not written from a place of judgment; I have no interest in shaming individuals working in the industry for their role in it. Everyone has to work somewhere, and as long as the structure is flawed, then it's impossible to avoid being complicit in some way. Of course, there are degrees of complicity, and we should choose ethically where we can, but merely to exist within the system means accepting some degree of compromise. In any case, everyone has their own moral compass, and their own breaking point.

I used to feel a lot of guilt about my place within the system, because I didn't know how else to cope with the systemic cruelty I was seeing in the world around me. It was a paralysing kind of guilt, the sort that made me want to shut out the world and flee inward, seeking refuge and atonement. But I knew that seeking refuge wasn't a solution, and eventually that guilt transformed into resolve — the kind that surges outward, determined to change the situation that brought on the guilt in the first place. No point in lamenting how things should have been; we can't change the past, and perhaps we had to go through the path we did in order to imagine alternatives. The best we can do now is look forward and chart a path toward a better future.

For a long time, I was uneasy about writing this book. I worried that people in the industry would react with contempt, casting me as a delusional SJW who was just bitter because I failed: my insignificant startup got rejected from Y Combinator and shut down, and now I was projecting my naive resentment at the whole industry; my critique would be dismissed as rooted in envy, not reason. As much as I now see the blind spots in that kind of reasoning, the possibility of it still disturbs me, because on some deep level it's how I judge myself.

And yes, it's true that if I had succeeded — if my startup had sold for an outrageous sum, and I had subsequently found myself in a comfortable sanctuary, able to ignore the rest of the world — I would not have written this book. I would have continued to swallow the reasoning of those who believe their success is deserved, dismissing any criticism of the terms of my success as motivated by jealousy. I can't blame those who might dismiss my critique now.

But I'm not writing this book for them. They can continue to believe whatever self-serving tales the powerful always tell themselves — and whisper to the power-hungry acolytes below them. What might make sense to them isn't necessarily what's best for everyone else. This book is meant for those whose belief has started to evaporate, and who are now thirsting for a narrative that speaks to their disillusionment. I write for those who are currently not in power, in the hopes that they'll see the world differently, and from there go on to be part of something I could never have imagined on my own.

The tragedy of the human condition is that we live in a world handed down by those who came before us. We learn it through the signposts they left for us to follow, and so our initial understandings are always flawed for reasons outside our control.

It's hard to make change if we take the world as a given. And yet we ourselves are part of this world, shaped by the same forces that we criticise. There's a bit of the villain within all of us, too. How do we make a world that is better, when this is the only world we know?

As ready as I am to fight for a better world, in some ways I can't even really imagine one; I can't unlearn a lifetime of passive ideological moulding overnight. My instinctive judgments reek of capitalist ideology; I've internalised capitalism's claim that those who make less money are less worthy and therefore deserve to suffer. Rationally I know this is absurd, but it means I can't fully trust my instincts, because my instincts are petty and narcissistic.

I haven't come across any guides for how to deal with this. I got here by piecing together ideas that rang true, following a silken thread of understanding laced with bolts of serendipity. Even as I've developed new axioms to dislodge the old, I don't know how to swap myself out for something new. Maybe I'll always carry bits and pieces of the old me, even the parts I hate.

In a way, this is liberating, because it relieves the pressure to be perfect. It's impossible to transform into the perfect person overnight — the person who is the right subject for the world you want to create. After all, you don't live in that world yet, and you're unlikely to get there anytime soon. Instead, you can focus outward, making the structural changes needed to ensure that the people who come after you are better people than you, living in a better world than you do now.

That brings us back to the beginning. What does it mean to abolish Silicon Valley?

Truth be told, the catchphrase began as a Twitter joke, one that adorned several other types of media before gracing the cover of the book you're now holding. I suspect the slogan resonated because people are angry, because they see the greed and waste in Silicon Valley as the vanguard of deeper problems in capitalism. Capitalism is a disease, and the venality of Silicon Valley is a morbid symptom, the fingers turning ashen. It's not the root of the problems, and it's not even the originator of many problems, but it magnifies them and makes them increasingly difficult to deny.

Abolishing Silicon Valley, then, means moving beyond the flawed paradigm of capitalism. It is irresponsible to allow technology development to be driven primarily by the needs of capital. Instead, we should rewire the lines of power within the industry, as well as society at large; the goal should be democratic control over technology's development, and an equitable distribution of its benefits.

From today's vantage point, in our barren landscape of creeping privatisation and billionaires amidst bone-chilling poverty, the idea of democratic control over technology seems almost unbearably far away. A murky fog of capitalist realism pervades, and the suffocating logic of commodification feels totalising, eternal. And yet even in the darkness of our dystopia there are glimmers of hope — glimpses of the world as it could be. Even in the heart of Silicon Valley itself, people are rejecting the status quo and organising to build power using whatever leverage they can find. As these pockets of resistance grow, they will shift the balance of power, charting a way out of our current conjuncture. What once seemed impossible will soon become inevitable, and maybe we'll wake up from this nightmare after all.

NOTES

Zero: Prologue

1 The term Luddite, while nowadays taken to mean "opponent of technological progress", actually has its origins in a much more nuanced historical movement. While the Luddites are often caricatured as prone to bursts of irrational tech hatred, situating their story in the proper political and economic context reveals their actions to be part of a long tradition of resistance by workers who have to fight for themselves because no one else will. For more on this, see Eric Hobsbawm's 1952 article "The Machine Breakers", or E.P. Thompson's *The Making of the English Working Class* (Penguin, 1992).

2 In 2018, Mark Zuckerberg testified in front of various Senate committees in the wake of the Cambridge Analytica revelations. The testimony was unsatisfying and unedifying, but the robot memes were amusing.

3 In February 2017, Bloomberg reported on a video by Uber driver Fawzi Kamel who happened to get then-CEO Travis Kalanick as a passenger. When Kamel explained that Uber's recent cuts to driver payments had hurt him financially, Kalanick replied that "some people don't like to take responsibility for their own shit". The video was released at a time of public relations turmoil for the company, and four months later, Kalanick would resign from his role. See "In Video, Uber CEO Argues With Driver Over Falling Fares" by Eric Newcomer for *Bloomberg*, published February 28, 2017, at https://www.bloomberg.com/news/articles/2017-02-28/in-video-uber-ceo-argues-with-driver-over-falling-fares.

4 For more on this concept, see Nancy Fraser's 2014 article "Behind Marx's Hidden Abode" in the *New Left Review*, or the book she co-authored with Rahel Jaeggi, *Capitalism: A Conversation in Critical Theory* (Polity, 2018).
5 My understanding of capitalism as a system of distinct classes builds on the theories first developed by Karl Marx, most famously elucidated in his 1867 magnum opus *Das Kapital* and later expanded by generations of Marxist thinkers.

One: No Girls on the Internet

1 James Damore's infamous memo was first reported by *Motherboard* on August 4, 2017, at https://www.vice.com/en_us/article/kzbm4a/employees-anti-diversity-manifesto-goes-internally-viral-at-google; the full ten-page document was published by *Gizmodo* the next day, at https://gizmodo.com/exclusive-heres-the-full-10-page-anti-diversity-screed-1797564320.
2 Joel Spolsky, CEO of Fog Creek Software and software blogger, wrote about his company's "red carpet" intern recruitment process for *Inc* in 2005. See https://www.inc.com/magazine/20070501/column-guest.html.
3 The O'Reilly Open Source Convention. I attended in 2011 and 2012.
4 The 2012 Quebec student movement — sometimes referred to as the "Maple Spring" — was a major student-led uprising sparked by the Quebec government announcing a 75% tuition hike at public universities.

Two: Googleyness

1 See https://abc.xyz/investor/founders-letters/2004-ipo-letter/.
2 TGIF originally stood for "Thank God It's Friday," and the events used to be held on Fridays, but they were moved to Thursdays not long before I joined in 2013, without changing the name.

3 In 2014, Google publicly released its diversity numbers for the first time. It was revealed that among technical roles, only 17% of employees were female, and that only 5% of its total US workforce were black or Hispanic. See "Google finally discloses its diversity record, and it's not good" by Murrey Jacobson for *PBS*, published May 28, 2014, at https://www.pbs.org/newshour/nation/google-discloses-workforce-diversity-data-good.

4 Google announced in March of 2013 that Reader would be retired on July 1. See https://googleblog.blogspot.com/2013/03/a-second-spring-of-cleaning.html.

5 In July of 2014, Google announced that it would be fully backtracking on their Real Names policy, no longer imposing restrictions on the names that users could choose for their Google+ accounts. See "Google reverses 'real names' policy, apologizes" by Violet Blue for *ZDNet*, published July 15, 2014, at https://www.zdnet.com/article/google-reverses-real-names-policy-apologizes/.

6 *Fortune* has listed Google as the #1 best company to work for in the US for every year between 2012 and 2017. See https://fortune.com/best-companies/2017/google/.

7 See https://www.vogue.com/article/hail-to-the-chief-yahoos-marissa-mayer.

Three: More than the Money

1 See http://www.paulgraham.com/gap.html.

Four: Fake It Till You Make It

1 A quote for a *Guardian* profile in December 2014. See https://www.theguardian.com/theobserver/2014/dec/21/travis-kalanick-uber-cab-app-observer-proifile.

2 The Canadian government runs a program called "SR&ED" (Scientific Research and Experimental Development) which

provides qualifying organisations with tax credits for wages and other expenses. The Quebec government offers supplemental funding.

3 "This CEO is Out for Blood" by Roger Parloff, published June 12, 2014, at https://fortune.com/2014/06/12/theranos-blood-holmes/.

4 The original graph is often attributed to Paul Graham, who is said to have drawn it on a whiteboard in collaboration with other Y Combinator partners during a private event for Y Combinator founders. See https://andrewchen.co/after-the-techcrunch-bump-life-in-the-trough-of-sorrow/.

5 "One Startup's Struggle to Survive Silicon Valley's Gold Rush", by Gideon Lewis-Kraus, first published April 22, 2014. See https://www.wired.com/2014/04/no-exit/.

6 In April 2014, Mark Zuckerberg announced that Facebook's infamous "Move Fast and Break Things" motto would be changed to the less catchy "Move Fast With Stable Infra". See https://mashable.com/2014/04/30/facebooks-new-mantra-move-fast-with-stability/.

Five: Accelerate

1 From a recorded interview for *VatorNews*. See https://vator.tv/news/2010-12-23-elon-musk-work-twice-as-hard-as-others.

2 A common (if vague) truism in the startup world is that 90% of startups fail. The truth is a little more complicated —- the real figures depend on how you define "failure" and "success", what time period you're considering, and what counts as a startup. For an analysis of the data, see "Conventional Wisdom Says 90% of Startups Fail. Data Says Otherwise." by Erin Griffith for *Fortune*, published June 27, 2017, at https://fortune.com/2017/06/27/startup-advice-data-failure/.

3 For a good explanation of the downsides of this paradigm, see "Toxic VC and the Marginal-Dollar Problem" by Eric Paley for *TechCrunch*, published October 26, 2017, at

https://techcrunch.com/2017/10/26/toxic-vc-and-the-marginal-dollar-problem/.

4 For more insight into why VC gravitates towards billion-dollar markets, see Scott Kupor's book *Secrets of Sand Hill Road* (Portfolio, 2019).

Six: Crash

1 From Peter Thiel's 2014 book *Zero to One: Notes on Startups, or How to Build the Future*, which is a condensed version of Thiel's lectures for an online class he taught on startups at Stanford University.

2 This statistic originated in an annual report released by the US Federal Reserve. The analysis of 2014, released in May 2015, reported that 47% of respondents "say they either could not cover an emergency expense costing $400, or would cover it by selling something or borrowing money". See https://www.federalreserve.gov/econresdata/2014-report-economic-well-being-us-households-201505.pdf.

Seven: Pivot

1 See https://www.businessinsider.com/jeff-bezos-net-worth-life-spending-2018-8.

2 Shkreli's company, Turing Pharmaceuticals, would later make the news for jacking up the prices of life-saving medication.

3 See https://nplusonemag.com/issue-25/on-the-fringe/uncanny-valley/.

4 See "World After Capital", by Union Square Ventures partner Albert Wenger. http://worldaftercapital.org.

Eight: Giving Up

1 Why do I have a quote from a lesser-known Kafka short story in a book about Silicon Valley? Good question. There's something

eerily timeless about Kafka, and when I came across these words (penned circa 1920), I was a little spooked at how precisely it captured the way I felt during the worst days of my startup.

2 See, for example, "Trump hosting roundtable with Silicon Valley's tech leaders" by Rebecca Shabad for *CBS News*, published December 14, 2016, at https://www.cbsnews.com/news/trump-tech-leaders-meeting-ibm-tim-cook-others/. The list of attendees included executives from Microsoft, Facebook, Google, Apple, Amazon, IBM, and Tesla.

3 See, for example, "Google Gets a Seat on the Trump Transition Team" by David Dayen for *The Intercept*, published November 15, 2016, at https://theintercept.com/2016/11/15/google-gets-a-seat-on-the-trump-transition-team/.

4 See "Y Combinator boss Sam Altman says he's not going to cut ties with Peter Thiel for supporting Donald Trump" by Peter Kafka (probably no relation) on October 16, 2016, for *Recode*, at https://www.vox.com/2016/10/16/13302120/y-combinator-sam-altman-peter-thiel-donald-trump. Thiel's affiliation with Y Combinator ended soon after that under unclear circumstances, as reported by Ryan Mac for *Buzzfeed News* on November 17, 2017: "Y Combinator Cuts Ties With Peter Thiel After Ending Part-Time Partner Program", at https://www.buzzfeednews.com/article/ryanmac/y-combinator-cuts-ties-with-peter-thiel-ends-part-time.

5 For a brief overview of incidents in this vein, see "The Ugly Unethical Underside of Silicon Valley" by Erin Griffith for *Fortune*, published December 28, 2016, at https://fortune.com/longform/silicon-valley-startups-fraud-venture-capital/.

6 See, for example, "Uber Drivers Speak Out: We're Making A Lot Less Money Than Uber Is Telling People" by Maya Kosoff for *Business Insider*, published October 29, 2014, at https://www.businessinsider.com/uber-drivers-say-theyre-making-less-than-minimum-wage-2014-10.

7 See, for example, "Inside an Amazon Warehouse, the Relentless Need to '"Make Rate"'" by Hamilton Nolan for *Gawker*,

published June 6, 2016, at https://gawker.com/inside-an-amazon-warehouse-the-relentless-need-to-mak-1780800336.

8 See, for example, "Foxconn Working Conditions Slammed bBy Workers Rights Group" by Steven Musil for *CNET*, published May 30, 2012, at https://www.cnet.com/news/foxconn-working-conditions-slammed-by-workers-rights-group/.

9 Several tech billionaires have signed "The Giving Pledge", a movement led by Bill Gates and Warren Buffet. These billionaires are essentially making a non-binding commitment to give away most of their ill-begotten wealth to philanthropic causes as opposed to, I don't know, spending it on private jets or burning it or whatever. It's not the worst thing they could do, but it's not exactly the pinnacle of sacrifice that it's often made out to be. See Anand Giridharadas' book *Winners Take All: The Elite Charade of Changing the World* (Allen Lane, 2018) for a critique of this sort of philanthropy.

10 Verso, 2011.

11 For coverage of this incident, see "Why GitHub's CEO Ditched Its Divisive 'Meritocracy' Rug" by Lauren Orsini for *ReadWrite*, published January 24, 2014, at https://readwrite.com/2014/01/24/github-meritocracy-rug/.

12 For a different but related take on this topic, see this interview with French economist Thomas Piketty for the International Inequalities Institute in July 2015, about his book *Capital in the Twenty-First Century* (you could also read the book, but it's a lot longer): https://medium.com/@LSEInequalities/an-interview-with-thomas-piketty-a972015438e0.

13 For coverage of this incident, see, for example, "Why Nancy Pelosi's Comments About Capitalism Disappointed Progressives" by Daniel Marans for *Huffington Post*, published February 1, 2017, at https://www.huffpost.com/entry/nancy-pelosi-town-hall-capitalism_n_58925a53e4b070cf8b807e28.

14 For example, Coca-Cola spent $4 billion on advertising in 2015, according to the FAQs page of their website, https://www.coca-colacompany.com/contact-us/faqs.

Nine: Profile Before You Optimise

1 As reported in John Carreyrou's book, *Bad Blood: Secrets and Lies in a Silicon Valley Startup* (Knopf, 2018).

2 For coverage of this fairly widespread phenomenon, see, for example, "Delivery Workers Are Being Cheated Out of Tips by Their Own Companies. This Isn't New." by Rebecca Jennings for *Vox*, published July 22, 2019, at https://www.vox.com/the-goods/2019/7/22/20703636/doordash-instacart-tip-policy.

3 For an excellent primer on UBI, see "The False Promise of Universal Basic Income" by Alyssa Battistoni for *Dissent Magazine's* Spring 2017 issue, at https://www.dissentmagazine.org/article/false-promise-universal-basic-income-andy-stern-ruger-bregman.

4 For a critical analysis of the gig economy, see *The Gig Economy: A Critical Introduction* by Jamie Graham Woodcock and Mark Graham (Polity, 2019).

5 For details on what it's like to work for a gig economy platform in the UK, see Callum Cant's book *Riding for Deliveroo: Resistance in the New Economy* (Polity, 2019).

6 The World Transformed, which began as a fringe festival for the UK Labour Party's annual convention, alternates between Brighton and Liverpool.

7 "Any industry that still has unions has potential energy that could be released by startups." Posted November 8, 2015, at https://twitter.com/paulg/status/663456748494127104.

8 The Historical Materialism conference takes place every November at SOAS University of London.

9 See, for example, "Contracts And Chaos: Inside Uber's Customer Service Struggles" by Johana Bhuiyan for *BuzzFeed News*, published March 6, 2016, at https://www.buzzfeednews.com/article/johanabhuiyan/contracts-and-chaos-inside-ubers-customer-service-struggles.

10 The problem isn't that tech companies are putting money towards efforts to expand computer literacy. The problem is the larger landscape which allows these tech companies to accrue excess wealth while public school systems are underfunded, and which also allows these companies to reap the benefits of their investment in the form of future lower wages. For commentary on large tech companies' investments in STEM education, see "The Tech Education Con" by J.S. Chen for *Jacobin*, https://jacobinmag.com/2019/01/stem-coding-bootcamp-education-scam-philanthropy, and "Tech's Push To Teach Coding Isn't About Kids' Success – It's About Cutting Wages" by Ben Tarnoff for the *Guardian*, https://www.theguardian.com/technology/2017/sep/21/coding-education-teaching-silicon-valley-wages.

11 I've posted a summary of this endeavour on my blog, at https://dellsystem.me/posts/a-year-of-200-books. Some of my favourites included Mark Fisher's *Capitalist Realism: Is There No Alternative?* (Zero Books, 2009) and Geoff Mann's *Disassembly Required: A Field Guide to Actually Existing Capitalism* (AK Press, 2013).

12 Harvard University Press, 2004.

13 Published by *Notes From Below* on January 29, 2018, at https://notesfrombelow.org/article/silicon-inquiry.

14 The workers, all US-based, later filed a complaint with the National Labor Relations Board for unfair dismissal, which would turn out successful. For an interview with some of the workers involved, see "Coding and Coercion" in *Jacobin*, published April 11, 2018, at https://www.jacobinmag.com/2018/04/lanetix-tech-workers-unionization-campaign-firing.

15 See, for example, "Microsoft Employees Protest Work With ICE, as Tech Industry Mobilizes Over Immigration" by Sheera Frenkel for the *New York Times*, published June 18, 2018, at https://www.nytimes.com/2018/06/19/technology/tech-companies-immigration-border.html.

16 In June of 2018, Google announced that they would not be renewing their contract with the U.S. Department of Defense. See, for example, “Google Plans Not to Renew Its Contract for Project Maven, a Controversial Pentagon Drone AI Imaging Program” by Kate Conger for *Gizmodo*, published June 1, 2018, at https://gizmodo.com/google-plans-not-to-renew-its-contract-for-project-mave-1826488620.

17 “Inside Google’s Shadow Workforce”, by Mark Bergen and Josh Eidelson, published July 25, 2018, at https://www.bloomberg.com/news/articles/2018-07-25/inside-google-s-shadow-workforce.

18 For details, see Eric Blanc’s book *Red State Revolt: The Teachers’ Strike Wave and Working-Class Politics* (Verso, 2019).

19 For details on the demands of the walkout, see the letter posted by the core organisers of the walkout at *The Cut*, at https://www.thecut.com/2018/11/google-walkout-organizers-explain-demands.html.

20 See “How Google Protected Andy Rubin, the ‘Father of Android’”, by Daisuke Wakabayashi and Katie Benner, for the *New York Times*, published October 25, 2018, at. https://www.nytimes.com/2018/10/25/technology/google-sexual-harassment-andy-rubin.html.

21 For coverage of the strike’s demands, see, for example, “Marriott workers just ended the largest hotel strike in US history” by Alexia Fernández Campbell for *Vox*, published December 4, 2018, at https://www.vox.com/policy-and-politics/2018/12/4/18125505/marriott-workers-end-strike-wage-raise.

Ten: Stupid Environment

1 Quoted from an interview with Adam Fisher for Fisher’s book on the history of Silicon Valley, *Valley of Genius: The Uncensored History of Silicon Valley (As Told by the Hackers, Founders, and Freaks Who Made It Boom)* (Twelve, 2018).

2 For more on this, see David Harvey's book *A Brief History of Neoliberalism* (Oxford University Press, 2007).
3 For a Marxist perspective on the economic underpinnings of the dotcom bubble and burst, see Nick Srnicek's *Platform Capitalism* (Polity, 2016). For insight into the psychological aspects of the bubble, Mel and Patricia Krantzler's *Down and Out in Silicon Valley: The High Cost of the High Tech Dream* (Prometheus, 2002) is excellent reading.
4 For commentary on this phenomenon, see, for example, "The iPhone X Proves How Tim Cook's Apple Is Embracing the Rich" by Paris Marx for *The Bold Italic*, published September 25, 2017, at https://thebolditalic.com/the-iphone-x-proves-how-tim-cooks-apple-is-embracing-the-rich-8a0270d8fc91.
5 OK, there are only two so far — Lime and Bird — but that's still two too many for my liking. See, for example, "Lime, A Scooter Startup That Barely Existed Two Years Ago, Now Going To Be Worth $2 billion" by Theodore Schleifer for *Vox*, published January 15, 2019, at https://www.vox.com/2019/1/15/18184756/lime-scooter-fundraising-valuation-two-billion.
6 I'm mostly referring to the SoftBank Vision Fund, which has poured incredible amounts of money into companies like the gloriously overvalued (and still unprofitable) WeWork, but SoftBank is only the most egregious example of the trend. For an overview of SoftBank's recent financial difficulties, see "SoftBank's Blurry Vision" by Alexander Salmon for *The American Prospect*, published October 11, 2019 at https://prospect.org/power/softbanks-blurry-vision-fund-tech-ipos/.
7 Scott Kapor's *Secrets of Sand Hill Road* (Portfolio, 2019) has a good breakdown of this, though an uncritical one.
8 The two American universities with the largest endowments are Harvard and Yale. Yale in particular has heavy exposure to venture capital through its fund, and the fund's manager, David Swensen, is the university's highest-paid employee, with a salary of $4.7 million in 2017. For a first-person account of

Yale University's attempts to prevent graduate students from unionising, see "Spadework" by Alyssa Battistoni in the Spring 2019 issue of *n+1*, published at https://nplusonemag.com/issue-34/politics/spadework/.

9 See, for example, "Purdue Pharma, maker of OxyContin, files for bankruptcy" by German Lopez for *Vox*, published September 16, 2019 at https://www.vox.com/policy-and-politics/2019/9/16/20868487/purdue-pharma-oxycontin-bankruptcy-opioid-epidemic.

10 See, for example, Emily Guendelsberger's book *On the Clock: What Low-Wage Work Did to Me and How It Drives America Insane* (Little, Brown and Co., 2019).

11 For commentary on corporation expansion in the era of digital Capitalism, see "Landlord 2.0: Tech's New Rentier Capitalism" by Jathan Sadowski for *OneZero*, published April 4, 2019, at https://onezero.medium.com/landlord-2-0-techs-new-rentier-capitalism-a0bfe491b463.

12 There's a concept called "elite projection" that neatly summarises this phenomenon. Public transit consultant Jarrett Walker explains it in a blog post called "The Dangers of Elite Projection", published July 31, 2017, at https://humantransit.org/2017/07/the-dangers-of-elite-projection.html.

13 See, for instance, Uber's introduction of its opaque "Safe Rides Fee" on every trip, which brought in nearly half a billion dollars of extra revenue for the company that was never specifically set aside for safety measures, as Mike Isaac reports in his book *Super Pumped: The Battle for Uber* (W. W. Norton, 2019).

14 See, for instance, "Tech Giants Amass a Lobbying Army for an Epic Washington Battle" by Cecilia Kang and Kenneth P. Vogel for the *New York Times*, published June 5, 2019 at https://www.nytimes.com/2019/06/05/us/politics/amazon-apple-facebook-google-lobbying.html.

15 For an overview of the history of financialisation, see Grace Blakeley's book *Stolen: How to Save the World from Financialisation* (Repeater, 2019).

16 See "The Next Recession Will Destroy Millennials" by Annie Lowrey, published August 26, 2019 at https://www.theatlantic.com/ideas/archive/2019/08/millennials-are-screwed-recession/596728/.

17 As reported by the real estate firm Compass for Q3 of 2019. See https://www.bayareamarketreports.com/trend/san-francisco-home-prices-market-trends-news.

18 See, for example, "SF's Homeless Count Reaches More Than 9,700" by Adam Brinklow for *Curbed*, published July 8, 2019, at https://sf.curbed.com/2019/7/8/20686653/san-francisco-sf-homeless-count-number-population-2019.

19 For details on San Francisco's affordable housing program, see "A Guide to Below-Market-Rate Housing in San Francisco" by Chris Roberts for *Curbed*, published October 8, 2018, at https://sf.curbed.com/2018/10/8/17902408/guide-bmr-affordable-housing-lottery-san-francisco.

20 For more on the difference between housing's status as exchange value versus use value, see David Madden and Peter Marcuse's book, *In Defense of Housing* (Verso, 2016).

21 See, for example, Anna Minton's *Big Capital: Who Is London For?* (Penguin, 2017).

22 For coverage of the unionising efforts at Anchor Brewing Co., see, for example, "The Brewers Who Make Iconic Anchor Steam Beer in S.F. Join Union" by Miranda Leitsinger for *KQED News*, published March 13, 2019, at https://www.kqed.org/news/11732763/the-brewers-who-make-iconic-anchor-steam-beer-in-s-f-join-union.

23 See https://onezero.medium.com/an-open-letter-to-uber-we-need-to-do-right-by-our-drivers-81453fad41e1.

24 See "Uber Lays Off 400 as Profitability Doubts Linger After I.P.O." by Kate Conger for the *New York Times*, published July 29, 2019, at https://www.nytimes.com/2019/07/29/technology/uber-job-cuts.html.

25 See "Uber lays off 435 people across engineering and product teams" by Megan Rose Dickey for *TechCrunch*, published

September 10, 2019, at https://techcrunch.com/2019/09/10/uber-lays-off-435-people-across-engineering-and-product-teams/.

26 See "Lyft Buys the Biggest Bike-Sharing Company in the US" by Sean O'Kane for *The Verge*, published July 2, 2018, at https://www.theverge.com/2018/7/2/17526892/lyft-buys-motivate-bike-sharing-expansion.

27 See "Uber Looked at Buying Bike-Sharing Firm Motivate before Lyft — But Walked Away Over Contract and Union Issues" by Sara Salinas for *CNBC*, published July 3, 2018, at https://www.cnbc.com/2018/07/03/uber-looked-at-motivate-before-lyft-contract-union-issues.html.

Eleven: A New Industrial Model

1 See https://www.businessinsider.com/sam-altman-world-with-trillionaires-is-inevitable-2016-6.

2 In 1976, manufacturing workers at Lucas Aerospace in the UK put forward a plan to direct their plant's output away from military contracts and towards more socially useful endeavours like medical equipment and clean energy, while also increasing the amount of worker control over production. This plan was suggested at a time when the company was 'restructuring' and was looking to lay off workers in the face of declining revenue. Ultimately, the workers' plan was rejected by management, but the proposals they put forward remain relevant today. For details, see https://lucasplan.org.uk/story-of-the-lucas-plan/.

3 See, for example, "CEOs Made 287 Times More Money Last Year Than Their Workers Did" by Alexia Fernández Campbell for *Vox*, published June 26, 2019 at https://www.vox.com/policy-and-politics/2019/6/26/18744304/ceo-pay-ratio-disclosure-2018.

4 For more information, see this page at *FoundSF*: http://www.foundsf.org/index.php?title=The_Hiring_Hall.

5 See https://labour.org.uk/wp-content/uploads/2018/09/Democratic-public-ownership-consulation.pdf.

6 The original report can be found at https://www.ucl.ac.uk/bartlett/igp/sites/bartlett/files/universal_basic_services_-_the_institute_for_global_prosperity_.pdf. For a critique of the limitations of the report, see "Universal Basic Services Won't Fix Our Economy" by Grace Blakeley for *New Socialist*, published February 8, 2018 at https://newsocialist.org.uk/universal-basic-services-wont-fix-our-economy/.

7 To read the case for scooters as a public service, see "Privately Owned Scooter Companies Don't Have a Future" by Paris Marx for *Jacobin*, published July 29, 2019, at https://www.jacobinmag.com/2019/07/e-scooters-bird-lime-uber-venture-capital.

8 See https://common-wealth.co.uk/bdc.html.

9 See, for instance, "The Unspeakable Cost of Parenthood" by Katherine Zoepf for the *New York Times*, published August 27, 2019, at https://parenting.nytimes.com/work-money/parents-money-stress. For some unknown reason, the Airbnb public policy team decided this story reflected well on Airbnb (rather than being a horrific indictment of the current housing landscape) and paid to promote the article through their @AirbnbCitizen account at https://twitter.com/AirbnbCitizen/status/1172272152449626114.

10 See "Ex-Google and Uber Engineer Anthony Levandowski Charged with Trade Secret Theft" by Andrew J. Hawkins for *The Verge*, published August 27, 2019 at https://www.theverge.com/2019/8/27/20835368/google-uber-engineer-trade-theft-secrets-anthony-levandowski-charged.

11 See "Reclaiming the Commons" by David Bollier for *Boston Review*, published June 1, 2002 at http://bostonreview.net/forum/david-bollier-reclaiming-commons. For an essay that compares today's tech giants to private American train operators in the mid-nineteenth century, see "Union Station"

by David A. Banks for *Real Life*, published September 3, 2019, at https://reallifemag.com/union-station/.

12 For a compelling argument in this vein, see "To Break Google's Monopoly on Search, Make Its Index Public" by Robert Epstein for *Bloomberg Businessweek*, published on July 15, 2019 at https://www.bloomberg.com/news/articles/2019-07-15/to-break-google-s-monopoly-on-search-make-its-index-public.

13 The web crawler research that eventually led to Google's founding was funded by National Science Foundation grants. See https://www.nsf.gov/discoveries/disc_summ.jsp?cntn_id=100660.

14 See https://medium.com/transit-app/welcome-to-scootopia-we-now-aggregate-all-electric-scooters-c31c7337d5e6.

15 I came across this suggestion in an interview with Aza Raskin, who co-founded the Center for Humane Technology, in issue 20 of *Offscreen*.

16 See, for example, "The Making of a YouTube Radical" by Kevin Roose for the *New York Times*, published June 8, 2019 at https://www.nytimes.com/interactive/2019/06/08/technology/youtube-radical.html.

17 See "Barcelona is Leading the Fightback Against Smart City Surveillance" by Thomas Graham for *WIRED*, published May 18, 2018 at https://www.wired.co.uk/article/barcelona-decidim-ada-colau-francesca-bria-decode.

18 Senator Warren published details of the plan via *Medium* on March 8, 2019, at https://medium.com/@teamwarren/heres-how-we-can-break-up-big-tech-9ad9e0da324c.

19 Although the underlying protocols are open, and although web browsers typically rely on open source technology, most browser usage falls under the purview of private corporations who are able to build moats through integrations with their other products. Google's Chrome browser alone commands more than half of all browser usage; Apple's Safari is the nearest competitor. For an analysis of this situation, see "Google Chrome Is Poised to Swallow the Whole Internet" by Eric Limer

for *Popular Mechanics*, published December 4, 2018 at https://www.popularmechanics.com/technology/infrastructure/a25400060/report-microsoft-kills-edge-google-chrome/.

20 For more on this, see Mark Fisher's *k-punk* blog post about the London 2012 Olympic Games, at http://k-punk.abstractdynamics.org/archives/011918.html.

21 See, for example, "Where Will the Ad versus Ad Blocker Arms Race End?" by Chris Baraniuk for *Scientific American*, published May 31, 2018 at https://www.scientificamerican.com/article/where-will-the-ad-versus-ad-blocker-arms-race-end/.

22 See "Building a Socialist Media System" by Leo Watkins, Tom Mills, and Dan Hind for *New Socialist*, published September 18, 2019 at https://newsocialist.org.uk/building-socialist-media-system/.

23 See "Game Developers Need to Unionize" by Tim Colwill for *Polygon*, published January 16, 2019, at https://www.polygon.com/2019/1/16/18178332/game-developer-union-crunch.

24 As an example, see French game studio Motion Twin, which describes itself as an anarcho-syndical workers cooperative.

Twelve: Epilogue

1 The quote is transcribed from a video interview with Ruairí McKiernan, published January 12, 2010 at https://www.ruairimckiernan.com/articles1/my-interview-with-internet-activist-aaron-swartz-1986-2013.

ACKNOWLEDGEMENTS

The first acknowledgement has to be to Repeater Books, for making this book come to life. Tariq, I still don't know what made you take a chance on me with a book deal, but I'm grateful nonetheless, and I'm glad that the book I ended up writing so closely mirrored the book you had in mind. Thanks also to Carl Neville for initially recommending me, Josh Turner for copyediting, Rhian E. Jones for proofreading, and the rest of the Repeater team.

Next, a massive thanks to my pseudonymised former co-founders. I'm grateful to have had the chance to build something from the ground up with you, and I've learned so much from our time together, which were some of the most rewarding years of my life — even if it wasn't always smooth sailing. Thank you all for reading this book so quickly, and for your thoughtful feedback. And for one co-founder in particular: I'll be forever grateful that we drifted leftward at the same time, even if our trajectories took us to different places. I couldn't have gotten here without you.

In my short time in London, I was lucky enough to get involved with several groups that were key to my political transformation. I'm grateful for my editorial colleagues at *New Socialist* and *Notes From Below*, my fellow activists in London Young Labour and Battersea Labour Party, and the students and lecturers in the International Inequalities Institute, especially my dissertation supervisor Bev Skeggs. Thank you all for showing me that there is a world outside the tech industry—a world populated by incredible people pursuing worthy goals. Particular thanks go to the

Notes From Below editorial crew for providing structure and context to my incipient interest in tech worker organising, and to Marijam Didžgalvytė for curating the technology issue with me.

Another major source of political education has been the reading group I started in London in early 2018. Thanks to all the attendees who contributed to the discussions, and thanks to the speakers who have been kind enough to share their work with a bunch of LSE students (and others) in a pub. Special thanks to speakers Grace Blakeley, David Madden, Will Stronge, Tom Mills, Nick Srnicek, Oonagh Ryder, David Adler, Amelia Horgan, Brett Scott, Jo Littler, Sahil Dutta, and Richard Seymour, whose talks influenced the arguments in this book. And much love goes out to Hettie O'Brien for taking over when I realised it was too much work for one person.

A profound thanks to all the editors who have commissioned me to write about politics and tech over the last couple of years, with some particular notables. Ben Tarnoff, you helped develop my inchoate thoughts on open source into an essay I could be proud of for *Logic* magazine, and your feedback on this book has been invaluable. Jamie Woodcock, you took a chance on me when you asked me to write "Silicon Inquiry" for *Notes From Below*; the ensuing essay paved the way for this book, which has also benefited greatly from your feedback. Ronan Burtenshaw, you commissioned me for a piece with the same title as this book for the first issue of the re-launched *Tribune*; your faith in me is and always has been gratifying, and I hope I'll be worthy of it.

Speaking of other media with the same title as this book: thanks to Riley Quinn of the podcast Trashfuture and Michael Walker of Novara Media, who both independently chose "Abolish Silicon Valley" to title episodes on which I guest-appeared on their respective platforms. Thanks also to everyone else who has interviewed me for a podcast, especially Dan Hind and Tom Mills of MediaDemocracy, Alex Doherty of Politics Theory Other, and Mat Lawrence formerly of IPPR. And thanks to Francesca Bria and Evgeny Morozov for inviting me to speak at the 2018 DECODE Symposium in Barcelona.

Thanks to anyone else who has played a role in the making of this book. This book has benefited greatly from conversations with Abi Ramanan, Ares Geovanos, Gifford Hartman, Jesse Squires, Moira Weigel, and Xavier Denis, all of whose experiences and perspectives have influenced my thinking. I'm very grateful for those who generously took the time to read an early draft, including Agnes Nemeth, Andrew Kortina, Chandler Abraham, Chloe Lim, Danny Spitzberg, Jimmy Wu, Matt Dannenberg, Paris Marx, Saining Li, and Shimmy Li. And Kulsoom Jafri, even though we've only known each for a few short years, and even though we've somehow ended up on separate continents, you're still a huge source of light in my life, and your dedication to the cause never fails to inspire me.

Thanks to the Tech Workers Coalition Writing Club for providing a much-needed occasion to shut up and write with likeminded people, as well as the San Francisco chapter of Shut Up and Write!, for more explicitly doing the same on Tuesday mornings in the Mission. And thanks to the Telepath community, Marc Bodnick and Tatiana Estévez in particular, for providing a safe and thoughtful space in which I was able to discuss ideas that made their way into this book.

A strange and ambivalent thanks to Paul Graham, whose essays once gave me hope that there would be a life for me after high school, and whose startup accelerator gave me the courage to embark on a path that would define the next several years of my life. Thanks, too, for continuing to share your opinions on controversial subjects like unions, inequality, and women in tech, inadvertently allowing me to realise that success in one area does not necessarily translate to universal wisdom, and that killing your heroes is an important part of growing up. And finally, thanks for unwittingly inspiring me to write my first piece of tech criticism by blocking me on Twitter — the resulting essay clearly rang a bell with others who had been blocked by you, who helped it go viral, and the overwhelmingly positive responses gave me the confidence to keep writing. This book could not have been written without you.

And finally, an unequivocal thanks to my other half: for putting up with my all-nighters and half-closed lids during the frenetic last weeks of writing this book, and for tolerating my ceaseless tirades about the tech industry even as you boarded the shuttle to Menlo Park every day so that we could continue to have health insurance. You've taught me so much in the short time we've known each other, and I'm so grateful to have you in my life, both as a life partner and as a source of political inspiration. Thank you Jason.

Repeater Books

is dedicated to the creation of a new reality. The landscape of twenty-first-century arts and letters is faded and inert, riven by fashionable cynicism, egotistical self-reference and a nostalgia for the recent past. Repeater intends to add its voice to those movements that wish to enter history and assert control over its currents, gathering together scattered and isolated voices with those who have already called for an escape from Capitalist Realism. Our desire is to publish in every sphere and genre, combining vigorous dissent and a pragmatic willingness to succeed where messianic abstraction and quiescent co-option have stalled: abstention is not an option: we are alive and we don't agree.